Preventive Maintenance
of Buildings

Preventive Maintenance of Buildings

Edited by

Raymond C. Matulionis

Department of Engineering Professional Development
University of Wisconsin-Madison

Joan C. Freitag

VNR VAN NOSTRAND REINHOLD
——————————— New York

Printed in the United States of America

Van Nostrand Reinhold
115 Fifth Avenue
New York, New York 10003

Chapman and Hall
2-6 Boundary Row
London, SE1 8HN, England

Thomas Nelson Australia
102 Dodds Street
South Melbourne 3205
Victoria, Australia

Nelson Canada
1120 Birchmount Road
Scarborough, Ontario MIK 5G4, Canada

16 15 14 13 12 11 10 9 8 7 6 5 4 3 2 1

Library of Congress Cataloging-in-Publication Data
Preventive maintenance of buildings / Raymond C. Matulionis, editor;
  Joan C. Freitag, associate editor.
      p.   cm.
  Includes bibliographical references.
  ISBN 0–442–31866–9
  1. Buildings—Protection. 2. Buildings—Maintenance. 3. Grounds
  maintenance. I. Matulionis, Raymond C.   II. Freitag, Joan C.
  TH9039.P73   1990
  690'.24—dc20   90–12126

Contents

10. Grounds   267

Gerald R. Andrews

Preface

This book provides a wealth of basic and state-of-the art information plus practical techniques to assist persons responsible for the preventive maintenance of major building envelope systems. Its pages address many of the concerns shared today by building owners, designers, and maintenance personnel who need to know why building systems fail; how to achieve the performance requirements of building components; how good design, materials, and workmanship can contribute to preventing premature failures; and how effective preventive maintenance measures can ensure a dependable, failure-free system. Many of the chapters also discuss inspection methods, since knowing where to look for potential failures and how to identify them are vital steps in an effective preventive maintenance program. Each chapter of the book contains an up-to-date bibliography and a list of professional associations with information on the topics dealt with in the respective chapters.

The contributors to this book—designers, builders, researchers, and educators—are experts in their fields, Each contributor offers valuable knowledge and insights gained over many years of experience. My own experience as an architect and my understanding of the field through my work as a continuing education program director at the Department of Engineering Professional Development, University of Wisconsin–Madison, has enabled me to select "in-the-know" contributors who address the readers as colleagues; the information they present is down-to-earth and practical—a sharing from practitioner to practitioner.

This comprehensive volume will benefit building owners who are concerned about their building's appearance and functioning and about the economic advantages resulting from a well-planned, cost-effective preventive maintenance program. The book will also provide valuable information to building designers and builders who continually strive to achieve failure-free building systems. Of course, all buildings must be maintained regardless of how well they are designed and built. Thus, this book is also written for the maintenance personnel who must be on the alert to detect potential building failures and to take the necessary steps to avert premature deterioration of buildings.

The editors express their appreciation to Darrell Petska for his review and final editing of a number of chapters in the book; to Kurt Waage for preparing the drawings; and to Marie Gutzmer and Tammy Helbing for typing the manuscript.

1. Introduction

Raymond C. Matulionis

The objective of preventive maintenance (PM) is to prevent premature building component failures in building systems. Procedures intercepting the deterioration of building components are PM procedures. Certain PM activities also include minor repairs or replacement of building components. Tuckpointing of masonry, for example, is the repair of mortar joints and is considered PM for the masonry system. However, the replacement of a substantial section of masonry construction would be a major repair and would not be part of a PM program.

Buildings will endure their expected lifetime if adequately maintained, which is the main principle of PM. Buildings have been built and maintained for thousands of years. The existence of many structures that are functional many years after they are built attests to their capability to last if adequately maintained. But when buildings are not maintained, early failures of building components occur and these often cause failures of entire building systems.

In order to deal effectively with PM, construction and PM personnel need skills and knowledge of the changing technology in the construction industry. They need to evaluate construction materials and techniques on a regular and continuing basis. Many building failures today can be traced to the ignorance of construction techniques by construction and maintenance personnel and the neglect of PM by the building owners or managers.

Building owners and managers are often unaware of the full potential of an effectively structured and implemented PM programs. In many cases, they simply hope failures will not occur. A PM program involves spending money for retaining qualified staff, developing PM plans, and carrying out PM activities. These are costs that building managers would like to avoid since they might not realize the benefits of the PM program in hard dollars and cents. Many times, early failures of building components are not visible and do not cause building distress. Thus, building managers might not believe that a PM program will offer cost benefits to the organization. Managers also often think that warranties will take care of various building components. They do not know that most warranties require PM during the term of the warranties.

Stated bluntly, unless there is interference with the building's function, generally the building manager will not allocate funds to perform PM activities. It takes a farsighted manager to invest in a PM program before inconvenience or even downtime has been experienced. Yet experience has proven that most building systems, if they are to fully serve their function, cannot be installed and forgotten. All systems require periodic inspections and regular maintenance. Otherwise, it is likely that systems will start to degrade at an accelerated rate very shortly after installation.

The objective of a PM program is to prevent failures from occurring prematurely. Some examples follow:

- If maintained by recoating, wood surfaces will not develop defects prematurely.
- If repaired, mortar joints in masonry walls will not leak excessively. (If not repaired, failure of the masonry system will occur much earlier than if the joints have been repaired.
- If maintained, expansion joints will prevent water from entering the system as well as fulfilling their function of accommodating the building's movement. (If not maintained, the water entry will result in rapid deterioration of the system.)
- If periodically recoated, metal cappings over parapets and counterflashings will not corrode prematurely.

Short-term and minor deterioration of building components may not be very detrimental to building systems if the process is arrested before the damage becomes severe. Also, the process of arresting early deterioration is generally quite inexpensive compared to repairs or replacement costs of the failed system. PM is most effective when it deals with potential failures and prevents actual failures. The maintenance of the system, therefore, interrupts the progression of the degradation process.

Theoretically, any system that is well maintained can exist indefinitely. Figure 1-1 shows the rate of deterioration rapidly increases with time for an unmaintained system. Where systems are not maintained and repairs are made only after failures have occurred, the rate of deterioration of the system can be substantially retarded through repairs but at greater costs. These greater costs are caused by systems' component replacement, by the interruption of activities taking place in the building, and by damage caused to the building. The bottom dashed line in Figure 1-1 shows continuous maintenance taking place, not allowing failures to occur. The normal building deterioration process is interrupted by PM.

The frequency, the extent, and the costs of PM depend on many factors including the type of system, the quality of materials, the system's design, and

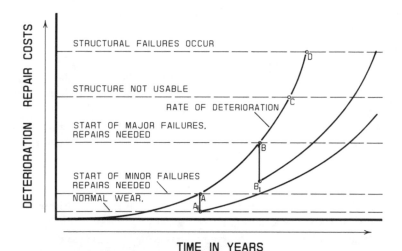

TIME IN YEARS

Figure 1-1. A curve showing deterioration rate and time relationship for maintained and unmaintained buildings. Points A through D represent stages of accelerated deterioration. Structures repaired at point A cost less overall and last longer than structures repaired at point B (compare curve A' to B'). *(Courtesy: Walker Parking Consultants/Engineers, Inc.)*

the installation techniques. Selecting an appropriate system is critical if the system is to perform effectively. Certain systems fail readily when exposed to extreme temperature fluctuations, excessive humidity, or the sun's ultraviolet rays. For example, masonry window sills and wall cappings will generally degrade rapidly if installed where moisture is present, especially if exposed to freezing temperatures. More appropriate components of this masonry system would be stone sills and stone cappings.

Performance of materials is dependent largely on their ability to withstand the elements to which they are exposed. For example, plywood installed close to grade will absorb moisture. Therefore, it is not an appropriate material for the situation. As you will learn in the forthcoming chapters, the key to successful PM of a building system is largely dependent on the quality of original construction. Materials selected and construction methods used significantly influence the degree of PM needed throughout the life of the building. The quality of building construction should be weighed against anticipated PM requirements. This process should be employed during the initial stages of project development. The initial costs of the system components and the costs of repair and replacement of these components become significant considerations when analyzing life cycle costs of the project. Inexpensive or cheap initial construction may turn out to be more costly over the lifetime of the system than if more expensive construction materials and techniques were used. Life cycle cost analysis is an invaluable tool when deciding what quality of building construction will be most cost effective. Material selection for the system should be based on performance requirements.

Knowledge of climate and other influences that degrade building systems is critical when detailing building construction. Construction detailing must reflect climatic needs. Improperly detailed systems may require constant maintenance, making costs of maintenance prohibitively high. Any system comprising the building envelope should be capable of shedding water. If allowed to penetrate construction, water can cause more damage than any other factor, especially if subjected to freezing and thawing. This case is true for paved areas, masonry walls, wood windows, and other building components. In the relatively dry regions, this factor does not call for a serious concern. But where substantial rainfall occurs, especially where freezing occurs, design against moisture should be a major requirement.

Quality of installation and workmanship of the system's components has elicited more complaints than any other aspect in building construction. Good materials and well-detailed construction without high-quality workmanship are not enough for ensuring efficient performance of building systems. Because of complexities inherent in construction, it is very difficult to ensure high-quality workmanship. Skilled workers and regular inspections of ongoing work are crucial to ensuring quality construction. These considerations are equally critical in PM of buildings.

When inspecting buildings for potential failures, decisions as to what type of repairs or maintenance the system requires often have to made on the spot. If the PM staff cannot quickly make a decision on the action to be taken, extensive damage to building envelope and building interiors may result.

PM of building envelope systems can be demanding, requiring well-developed skills of PM staff for inspecting and analyzing existing conditions (how urgent is the situation and how severe is the problem), and for taking necessary action without delay, especially if the failure is imminent. Once the problem is identified, maintenance and/or repair methods have to be selected. Individuals responsible for PM should know the process of maintenance and repair. They need to be familiar with the system's components and their functions. What causes or tends to cause failures of various components of the system should be understood by all PM staff.

The following basic influences play a role in degrading building systems: moisture, temperature, wind, contaminants, ultraviolet rays, poor design and construction, and mechanical damage. When dealing with the potential of moisture entry into the system, good drainage and designs that provide for water shedding eliminate many problems. Problems due to temperature fluctuations can be avoided by providing enough properly designed and properly installed building joints. Anchoring materials provide resistance to strong winds. Appropriate materials and coatings resist contaminant attacks, while shielding protects system components against ultraviolet rays. Poor design and construction can be overcome by better understanding of building construction methods and materials. Mechanical damage can be prevented by using appropriate protective devices.

As already mentioned, inspection helps ensure high-quality PM or repair work. Inspection, if it is to be effective, must be performed systematically. The inspector must be capable of reading signs of potential premature failures. The inspector must know where to look for potential problems and should be familiar with indications of the need for maintenance work or system repairs. Also, the amount of damage has to be assessed. In new construction, inspectors must make sure the work complies with construction documents (drawings and specifications). Besides inspecting maintenance or repair work, PM requires the inspector to find the problem, identify its severity, and in many cases, recommend the necessary procedures for solving the problem.

To prevent recurrence of the failure, the cause of degradation should be known. Decay of wood windows, for example, can be caused by moisture that enters the window through the joints of the window frame assembly, wetting areas within the assembly for lengthy periods of time. Exterior coatings of the window frame, in this case, would not improve the situation but make it worse by preventing the wood from drying out. Thus, the problem has be corrected before recoating the wood surfaces. Keeping records of as-built drawings and specifications can be most helpful in resolving this type of problem.

A good approach for starting PM of building systems is to rectify problems (if any) that are caused by faulty initial design and construction. Final inspection at the time of project completion generally resolves this aspect. But if there are areas that are susceptible to failure, modifications have to be made so that components in these areas are not exposed to above-normal degradation potential. If there are areas that appear to be susceptible to premature failures, these areas must be clearly marked and inspected continuously. If precautionary steps aren't taken, significant damage to the system could occur simultaneously with the failure of the component.

Any building program dealing with new construction or maintenance includes two basic parts: the process of work implementation and the management of the process. In PM, knowing how to perform the work needed to prevent the possibility of failures is only one part of the process. The program has to be effectively managed. Chapter 2 addresses this concern. It discusses advantages of effective PM programs, the process of setting objectives, and organization of the program itself. Chapter 2 also explores how much to invest in the PM program: It is critical to establish a break-even point for the PM program.

Generally, it is difficult to sell the program to managers of an organization. Although failures are always more expensive to correct than the performance of PM activities, managers are often reluctant to allocate sufficient funds for maintaining the buildings unless there are visible signs of imminent and costly failure. Chapter 2 also calls attention to this concern.

Of all building systems, roofing systems cause the greatest concern to the facilities managers because of the potential damage that failure can cause not only to the roof itself but to the interior finishes and equipment. Roofing failures can also disrupt the activities taking place in the building, as well as pose a physical danger to the workers and other people in the vicinity. Chapters 3, 4, and 5 deal with built-up roofs, single-ply roofs, and metal roofing systems, respectively.

Built-up roofs are proven traditional roofs that are used extensively today. If properly constructed and maintained, these roofs provide long and efficient service. PM of these roofs can be most cost-effective if performed correctly.

Single-ply roofs are relatively new systems. Over 50% of roofs installed today resort to this system. The requirements for installation, repair, and maintenance of single-ply roofs are very different from built-up roofs. Also, because these systems have not been in existence in the United States for much more than 10 years, not many construction professionals have an in-depth knowledge of how they function and what maintenance and repair procedures work. Because of changes in the single-ply systems' installation specifications, and changes in the formulation of membranes themselves, it is extremely difficult to keep up-to-date with the technology. Chapter 4 explains single-ply roofing systems and recommends PM procedures.

Metal roofs are another area that presents concerns and calls for knowledge and skill in design, installation, and maintenance. Today, metal roofs are used extensively on single-story nonresidential buildings. These roofs today can be subdivided into three categories: through-fastened roofs, architectural standing seam roofs, and structural standing seam roofs. Metal roofs require specialized skills in their potential failure analysis and PM, and these are dealt with in Chapter 5.

Concrete failures also concern facility managers. Concrete has been in use for many centuries. The use of concrete can be traced to its use by Romans for aqueducts. Some of the Roman structures are still in existence today, indicating their great knowledge of concrete. Some experts, however, believe that climate and the lack of contaminants in the area when these structures were erected were the real reasons for the longevity of Roman concrete.

Today, concrete failures are numerous. Spalling and cracking of concrete, rusting of reinforcing bars, and a variety of other impairments are not difficult to find. Moisture, freezing, road salt, and other influences cause deterioration of concrete construction. Chapter 6 analyzes reasons for concrete degradation and makes recommendations on how to maintain a trouble-free concrete.

PM of exterior wood surfaces is dealt with in Chapter 7. Surface preparation, whether for new paint or stain application, or for maintenance of wood surfaces, is critical. Understanding the substrate to which coatings are applied and formulations of coatings and stains are just as important.

Chapter 8 deals with concrete and clay masonry maintenance. Masonry construction has been in existence for thousands of years, yet there are numerous masonry building failures today. External moisture, interior humidity, gravity, pressure differentials between the outside and inside of the building, wind, faulty materials, and shoddy workmanship all contribute to masonry wall failures. Designers, builders, and PM staff can control these factors that make masonry walls the second most common cause for litigation on buildings.

Chapter 9 addresses curtain wall problems and suggests procedures for maintaining them. Curtain wall construction includes non-load-bearing brick or concrete masonry, or concrete and metal panels. Some of the failures and PM procedures relating to masonry and concrete construction are addressed in Chapters 6 and 8. Chapter 9 analyzes PM procedures applicable to metal curtain wall construction.

Chapter 10 deals with grounds maintenance. Although grounds maintenance may not seem as critical as the maintenance of building systems, having problem-free and usable grounds (lawns, drives, parking areas) can be critical, especially in a public building. Furthermore, sometimes the image of the property is portrayed more directly by the appearance of the grounds than the building itself. Grounds maintenance is a very complex process. It deals with a growing, ever-changing palette. It requires different considerations for different climates and seasons, knowledge of different chemicals, and knowledge of grounds maintenance equipment. Different plants require different soils, fertilization, watering schedules, and times for planting and pruning. Providing good drainage to lawns, drives, and parking areas is of paramount importance. Therefore, design of grounds should be handled with skill and care if PM problems are to be manageable.

In conclusion, this book provides guidelines for PM of major building systems and grounds. Hundreds of new materials for use in different building systems appear on the market yearly. Some of these have been rather short-lived for various reasons, including early failure and high costs.

This book does not deal with specific building products; it deals with principles of maintenance that have proven to be effective in extending the life of building systems. The chapters review materials and components of building systems, their properties, and their appropriateness for the use in buildings. Design, construction detailing, and installation considerations are reviewed. The chapters emphasize that original design and installation of building components play a critical role in the maintenance of building systems. Maintenance of poorly constructed systems is generally wasteful and extremely costly.

The chapters stress the importance of knowing the causes of building failures. The factors causing degradation of building systems should be understood by designers, construction personnel, and maintenance staff. This knowledge plays a critical role in selecting materials and techniques not only in new construction but also in maintenance. Inspection, a key element in PM, is underscored in this book. Identifying potential problems is emphasized here with illustrations and checklists. Finally, PM and minor repair procedures are outlined.

A well-maintained building system can provide many years of satisfactory service. Maintenance is often less expensive than repairs. Failures not only can cause distress to the system but can affect activities taking place within the building. Additionally, PM contributes to the morale of the users of the facility and enhances its image and value.

2. Organizing Preventive Maintenance Activities

Stephen R. Mulvihill

2.1 INTRODUCTION

An efficient preventive maintenance (PM) program for buildings brings numerous benefits to your company. You can apply PM procedures to the components of the building shell, HVAC (heating, ventilating, air-conditioning), electrical, plumbing, and other systems. If you monitor the performance of building systems and halt deterioration of the systems components before expected failure occurs, you realize substantial cost savings. In addition to the savings, a well-maintained facility provides a desirable work environment for employees and presents a good impression to the public.

The meaning of PM must be clearly understood before planning and implementing a PM program is discussed. Many facility managers believe a PM program mainly involves the systematic repairing of components that break down, which is true only to a degree. A more accurate and complete definition of a PM program is as follows: It is a systematic and routine process of preemptive inspections and minor repairs of building systems and equipment that ensure building systems and equipment performance to the set target dates. These dates are based on the desired life span of the building, the holding period of the building, or the life of the specific system.

PM activities are generated through regularly performed inspections and through established routine work on building components and equipment (Fig. 2-1). PM may also be initiated by minor failures of building systems. Repeated failures signal the need for a capital outlay for replacement or a minor repair. Once a need for PM work is identified, it should be described in the form of work orders (Fig. 2-2). These are then fed into the PM manager's information system for cost accounting, budgeting, and scheduling the PM activities. This information system also tracks and prints results of the management objectives and PM work accomplished. The system maintains records that demonstrate to the company management the benefits of the PM program.

In addition to defining PM, it is important to define building components

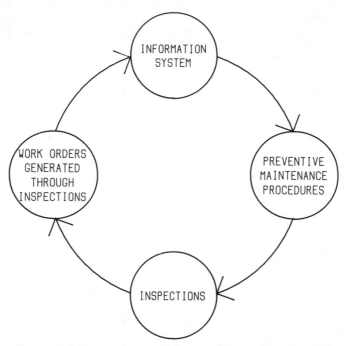

Figure 2-1. Preventive maintenance program elements.

and building systems. Building systems are made up of building components. Each component forms a link in the chain that can be construed to represent a system. A building system, therefore, is only as good as its weakest link. Each building component must be treated with equal importance in order to maintain the system's integrity. A built-up roof system, for example, is made up of components consisting of the structural roof deck, insulation, felts, bitumen, a surface layer of bituminous flood coat and gravel, and appropriate flashings. All of these components must be in good repair in order to keep water out of the system and the interior of the building.

The PM field is complex, requiring such activities as developing a reference library and record-keeping system, inspecting building components, designing and implementing procedures, maintaining qualified staff and adequate inventories, budgeting work, and implementing a safety program. Someone who has just assumed the management of the physical plant, including PM responsibilities, may be dismayed at the variety and complexity of tasks. Therefore organization and efficiency are imperative.

2.2 IMPORTANCE OF PREVENTIVE MAINTENANCE

Preventive maintenance of buildings provides many benefits to the users and owners of any facility. It stops or minimizes premature failures of building components, protecting the investment. Prevention of failures is in most cases drastically less expensive than repair of failures. Additionally, PM preserves a desirable image of the facility, which is an important feature for users' morale as well as for the sale of the building. A well-maintained building is also a safer building. These factors contribute significantly toward making the occupants of the facility more productive, whether in a residential environment, a recreational complex, or a workplace.

2.2.1 Extending the Life of Building Envelope Systems

The need for PM of mechanical and electrical equipment is much more apparent than the need for PM of building envelope systems. For example, there is a great concern for ensuring the elevators in the building are always operational, the heating and cooling systems work continuously, and the electrical service is uninterrupted. However, building envelope systems receive much less attention. How frequently, for example, are the masonry and the concrete surfaces inspected for potential failures? Lack of PM on building systems can shorten their life span or can result in extensive repairs or replacement of components at staggering costs to the company.

2.2.2 Selling the Building

Today's real estate purchasers are increasingly knowledgeable about the buildings they buy. They repeatedly engage consultants to assess conditions of building systems. Consequently, the owner can no longer hide an unmaintained building. If the owner invests inadequate effort in the upkeep of a building during a holding period, the market will adjust for the minimal maintenance effort with a lower sale price.

2.2.3 Accommodating Physical Changes of the Facility

Since expansion and modification of physical facilities are inevitable, a PM program can offer significant support for the building system in transition.

CENTRAL MAINTENANCE OFFICE USE

DATE RECEIVED: _____

JOB REVIEW DATE: _____

WORK CATEGORY: (CIRCLE)

CARPENTRY
ELECTRICAL
MECHANICAL
PIPE FITTING
PLUMBING
GROUNDS
PAVING
PAINTING
REFRIGERATION
TRUCKING
LOCKS/HARDWARE
ROOFING
MASONRY/EXTERIOR WALLS
CONCRETE
WINDOWS
OTHER

WE/THEY PLAN TO BEGIN WORK_____

WE/THEY PLAN TO FINISH WORK_____

HOLDING FOR: (CIRCLE)

INFORMATION
PARTS
LABOR NOT AVAILABLE
AVAILABILITY OF WORK AREA
APPROVAL BY ADMINISTRATION
SPEC BY CONSULTANT

ASSIGNED TO:

OUTSIDE CONTRACTOR_____
OTHER_____
MAINTENANCE_____
BUILDING MAINTENANCE_____

WORK REQUEST NUMBER:_____

COST INFORMATION:

MATERIAL_____
          _____MARK-UP
LABOR_____
          _____MARK-UP
PROPOSAL_____
OR
COST TOTAL_____

PERSONNEL	RATE	HOUR	DATE	EXTENDED AMOUNT

FORM 100

WHITE   - CM FILE (COMPLETION COPY)
GREEN   - INFORMATION COPY
CANARY   - CM PARTS
PINK   - CM LABOR
GOLDEN ROD - ORIGINATOR

12

WORK REQUEST

DATE: _____

BUILING: (CIRCLE)   PRIORITY: (CIRCLE)   REQUESTED BY: _____

WCHS   COMPLETED EMERGENCY _____

WNHS   EMERGENCY _____   ADMINISTRATOR

JHS   URGENT (1 WEEK) _____   APPROVAL: _____

TPMS   NECESSARY BUT CAN'T BE

EES   ACCOMPLISHED BY CENTRAL

RES   MAINTENANCE FORCE _____   DATE NEEDED: _____

        ACCOUNT TO BE

        CHARGED: _____

AREA OF ROOM: _____

DETAILED DESCRIPTION OF WORK (ATTACH SKETCH IF AVAILABLE)   SKETCH AREA

1. _____

2. _____

3. _____

4. _____

5. _____

6. _____

7. _____

8. _____

9. _____

10. _____

WORK COMPLETED & ACCEPTED

ADMINISTRATOR APPROVAL: _____   DATE: _____

Figure 2-2. A work order describing work to be done and providing a record of complete maintenance.

For example, PM personnel are often responsible for managing areas where asbestos and other hazardous substances are located. An established PM program can respond efficiently and economically to these and many unforeseen demands. If, on the other hand, a company does not have a PM program, some type of entity would invariably have to be structured, usually on short notice, to respond to these requirements. Remember, it takes time to develop an effective PM program. A group of employees, structured on short notice to respond to a specific need (unless the problem is highly specialized and easily definable), generally cannot be as effective as an efficiently functioning in-house maintenance team.

2.2.4   Maintaining a Desirable Image

A facility in a state of disrepair does not contribute positively to its employees' morale and does not impress visitors. The grounds, including lawns, vegetation, drives, walks, and parking areas, are observed first by visitors. Second, exterior surfaces of buildings contribute to the image employees and visitors form about the company. Finally, the appearance of interior spaces also influences opinions.

2.2.5   Reducing the Cost of a Product

The overall cost of any product or service has fixed as well as variable components. PM focuses on the fixed component by reducing the overall building maintenance budget, of which a percentage is factored into every "widget" the firm produces. It can also reduce other more subtle or unplanned-for overhead expenses. These expenses could be tied to safety, litigation, and environmental concerns.

2.2.6   Ensuring Employee Safety

Unmaintained buildings are unsafe. The personal injury potential resulting from a failure of building envelope systems is considerable. The roof with blocked drains allowing a deep buildup of water may cause a catastrophic collapse. This dangerous condition would be identified in a routine PM inspection. Water intrusion from continuing roof leaks can seriously damage the roof deck or building structural systems. Deteriorated and unmaintained bleachers and grandstands can collapse. Old ceiling systems in need of maintenance work can fail and cause injuries to the occupants of the building.

An effective PM program helps to prevent injuries by ensuring that buildings do not fail unexpectedly. Improved building safety through PM work may be somewhat difficult to measure; however, the numbers of insurance claims, accidents, or incidents of employee absenteeism can indicate its effectiveness.

2.3 SETTING PREVENTIVE MAINTENANCE OBJECTIVES

A PM program for any type of facility includes defining the process of work to be performed on the building systems. This process can be organized in terms of objectives to be achieved. One of the fundamental goals of a PM program is to have achievements documented through irrefutable, empirical data. Such data are essential to show results to management and to gather momentum for the program. Other objectives include defining the scope of the program, costs, management procedures, and the PM work itself. These objectives are common to all PM programs, although they assume individualized forms that respond to manager styles and specific needs of different facilities.

2.3.1 Defining the Scope of the Program

An effective PM program outlines areas to be maintained in order of importance and defines responsibilities of the PM manager and PM staff. In most cases, the responsibility is given to the facilities director, who is also responsible for custodial work, repairs, capital improvements, and energy conservation. This constitutes an assignment that is much too demanding for a single individual. For effective PM budgeting, training of PM personnel, monitoring building systems' performance, and the actual performing of work must be assigned to a separate individual—a competent PM manager.

The amount of PM investment versus the benefit can be measured in the cost of critical failure. Some features of a building are easily replaced and should not be maintained if they can be replaced with little downtime or cost. Small horse-powered motors are a good example. In contrast, fluorescent lamps are generally not replaced on a one-on-one basis but by entire banks or floors of fixtures as they start to fail. Systems on buildings scheduled to be demolished in the near future should receive minimum maintenance, except for safety purposes.

A premature building failure may result in damage to building contents, loss of production capacity, loss of the building, loss of market potential, or even loss of lives. Heating and cooling, electrical, fire protection systems, and some building shell systems are high-priority systems for maintenance. The critical building shell systems generally are roofs, exterior masonry surfaces, and foundations.

Systems such as roofs have many potential costs if failure occurs. Owners are often forced to shut down production in order to repair or replace a failed roof, which, of course, could be very costly. In addition to product sales loss, there are the overtime roofing labor costs required by an accelerated project. Also, to meet the urgent need, the owner may be forced into an undesirable roofing alternative that could be more expensive in the long run. Many owners have learned too late that some contemporary roofing systems, while appearing to lend themselves to the emergency situations, do not have the promoted life expectancy.

Structural systems of a building also require careful scrutiny. You should be alert for signs that signal possible structural failure. For example, retrofit suspended ceiling grid and lighting fixtures can be very dangerous if attached to the original ceiling rather than to the structural system. Proper installation is critical if the existing ceiling is heavyweight lath and plaster over wood frame construction. Beware of any subtle deflections in ceiling systems, since many old lath and plaster ceilings fail gradually over several years. The nails or staples holding the metal lath will pull out, risking a life-threatening ceiling collapse.

Another critical structure-related area to monitor is roof leaks. When a roof leak is stopped the possibility of structural damage still exists, particularly if the leak has persisted for many years. While it is generally known that gypsum and cementitious wood fiber decks are vulnerable to water damage, metal roof decks can also rust in about 5 years under wet roof insulation or recurring roof leaks. Failure of the entire roof system can result.

New mechanical equipment and piping installations in an existing building must be inspected frequently to identify any possible discrepancies that could have been caused to the structural system by the new construction. Careless mechanics sometimes cut the webs of bar joists to facilitate installation of piping. Piping or mechanical equipment is routinely, often improperly, hung from the roof construction. The only warning the PM inspector may get of an imminent structural failure is the sudden appearance of an unexplained roof pond. Careful inspection of the structure where construction has taken place is critical.

Swimming pools, cold storage buildings, any buildings with high interior humidity, and buildings with caustic production processes pose special threats to metal decks, metal structural connections, metal fire dampers, and metal fire doors. While buildings should be routinely maintained, PM work should be prioritized to be cost-effective. Focus on the building systems affecting the entire facility operation or those having high liability potential.

2.3.2 Identifying Costs

Probably the most fundamental objective of the PM program is to identify the pattern of costs connected with the maintenance of various building systems. A useful pattern of costs could take years to develop and requires effective record keeping of all maintenance work performed on the building. Building envelope systems generally take longer to deteriorate than mechanical or electrical systems. If a reliable pattern of costs is not available at the outset, the cost information should be obtained from PM work done on similar buildings.

PM of building systems is an investment. The investment consists of the funds for PM inventories, salaries, equipment, and the PM work itself. Realize that a PM program could represent a substantial capital investment for the company. Like any investment, it should result in a desirable return, with savings coming from decreased frequency of repairs, reduced premium time on repairs, reduced energy consumption, and decreased insurance expenditures.

Since PM significantly affects the total operational budget of the facility, and since a well-administered and adequately budgeted PM program results in a wise investment for the company, it is imperative to include the PM program in the long-range planning of facilities operations. Performance of the building systems should be adequately monitored so that reasonably accurate predictions can be made as to the amount of PM work needed on the systems.

2.3.3 Developing Effective Management Procedures

A qualified and skillfully managed PM work force can be an asset to any company. Unqualified workers following poorly structured PM procedures may exhibit a high level of activity as a result of short-notice responses to failures, but this approach is usually inefficient and costly. After instituting an effective PM program, an increased level of activity (representing PM work that has not been adequately performed before the program commenced) generally takes place. As the program progresses, the work level increases, but peaks and declines as there are fewer systems and equipment breakdowns (Fig. 2-3). At some point, the level crosses the original baseline and represents savings not realized before instituting an effective PM program. As a result of these savings, there is a potential to add further work orders for those projects that the staff did not have time for, or there is potential for physical plant operations cost reduction.

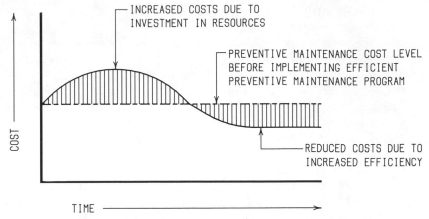

Figure 2-3. Manpower/resource management potential.

2.3.4   Ensuring Continuity of a Preventive Maintenance Program

For any PM program to be effective, it must have continuity: The performance of building systems and equipment must be monitored and documented on a continuous basis and PM must be performed without delay. This policy must extend throughout the life of the building if benefits of the PM program are to be realized. The maintenance records provide continuity and guidance.

Since it takes time to develop records of repairs and maintenance, record-keeping procedures must be implemented with the first PM task. The continuity of an effective PM program also depends on an organization's ability to retain a qualified work force and capable management. It is imperative that a PM program is effectively staffed at all times. If PM information on the building is not recorded, but is committed to the memories of certain individuals, the staff issue can be critical. Loss of key maintenance personnel can cause a significant void in the overall maintenance program. A well-documented and organized PM program can overcome this problem.

2.4   STRUCTURING A PREVENTIVE MAINTENANCE PROGRAM

For PM administrators initiating a PM program for their buildings there are well-defined necessary steps that can be followed. For those managers with established PM programs, a review of the steps may highlight the often overlooked, yet necessary, issues. The points to consider include personnel qualifications, critical areas of the facility, high-priority building systems, reference centers, inspection procedures, paperwork, information storage and retrieval systems, check points for monitoring the achievement of PM objectives, access to building systems and components, corrective repairs, and managing changes.

2.4.1 Supervisor/Manager Qualifications

The first and the most important step toward developing an effective PM program (or toward ensuring the continuity of an already functioning program) is the selection of a PM program manager. This individual should be experienced, responsible, and sufficiently aggressive to implement the program (or to continue administering the program if it is already in place). He or she could be responsible not only for the PM work on the building system but also for the PM work on mechanical and electrical equipment, which generally assumes the higher priority.

2.4.2 Identifying Critical Areas of the Facility

The next step is to survey the facility and identify those building areas prone to sustaining extensive and costly damage that could result in shutting down operations due to building system failure. Because of the increased consequential damages, these areas warrant a higher level of scrutiny and investment. Remember that consequential damages are never covered in construction or equipment guarantees and the owner assumes the risks. Consider the importance of the roof system sheltering key production equipment, the computer facility, or the president's office.

2.4.3 Inventorying High-Priority Building Systems

Next, start inventorying high-priority systems such as heating and electrical equipment and roofs. Identify, list, and tag them individually. Then include other systems of lower priority. There are many critical but overlooked systems throughout the building. The PM manager must identify these areas and familiarize himself with them. They include the fire alarm, sprinklers, and fire shutdown systems, as well as exits and self-closing doors and hatches. (These are mechanically operating devices such as closers, panic bars, etc.) In certain buildings with high pedestrian traffic, simple footing conditions on floor finishes is a critical issue. Consider the important features of bleachers and railing, such as structural integrity, especially for exterior installations.

```
ROOF HISTORY RECORD--ROOF TEST CUT ANALYSIS
_____

TEST CUT NO.  _____   PROJECT NO.  _____  DATE OF TEST CUT  _____

ROOF AREA NO. _____    MANUFACTURER/DATE/SPEC. NO./CONTRACTOR

SAMPLE SIZE   _____    PRIME ROOF:  _____
                        OVERROOF:    _____
BUILDING                             OWNER

_____        _____
                                     _____
YEAR BUILDING CONSTRUCTED _____     _____


         DESCRIPTION/SPEC. NO.       OBSERVATIONS

Overroof         _____   _____
Surfacing:       _____   _____

Overroof         _____   _____
Membrane:        _____   _____
Flashing:        _____   _____
Overroof         _____   _____
Insulation:      _____   _____

Prime Roof       _____   _____
Surfacing:       _____   _____

Prime Roof       _____   _____
Membrane:        _____   _____
Flashing:        _____   _____

Insulation       _____   _____
(top layer)      _____   _____
                 _____   _____

Insulation       _____   _____
(bottom          _____   _____
 layer)          _____   _____

Vapor            _____   _____
Barrier:         _____   _____

Roof Deck:       _____   _____
                 _____   _____

Comments:        _____
_____
_____

Guarantee Information: _____

Inspector:                           Testing Service:
_____        _____
```

Figure 2-4. Guide for identifying the condition of existing construction.

```
LOCATOR LOG

                         FIRM/ADDRESS/CONTACT/PHONE

ARCHITECT
GENERAL CONTRACTOR       _____
ROOFING CONTRACTOR       _____
ROOF CONSULTANT          _____
TESTING LABORATORY       _____
ROOFING MANUFACTURER     _____
DECK CONTRACTOR          _____
DECK MANUFACTURER        _____
SHEET METAL CONTR.       _____
PLUMBING CONTRACTOR      _____

_____


INSURANCE CLAIM INFORMATION

_____
_____
_____
_____
```

Figure 2-4. *(Continued)*

After identifying the building systems to periodically undergo PM, start gathering all available information on them. Retrieve all construction documents and shop drawings. These are often buried in someone's vault, since they may be the only copies in existence. If the management is reluctant in allowing these drawings to be used by the maintenance staff, invest in reproducible sets of these documents and retire the originals to the vault.

Where no documentation exists, the PM manager must initiate a detailed survey of construction and conditions of building systems, including roofs, walls, foundations, fenestration, entries, exits, critical interior surfaces, grounds, and of course, mechanical and electrical systems. Include scale drawings of the systems, a description of construction, information on designer and contractor, year of completion, and a list of remaining warranties. For such systems as roofing, secure actual core cuts of the existing construction (Fig. 2-4). For landscape planning, obtain soils analysis data. This data should be substantiated by any available documents. Obviously, the ideal situation would be to have as-built drawings and specifications. When doing field sampling be careful not to void guarantees. Any intrusion into a guaranteed roof can only be done by a manufacturer-approved contractor, or the guarantee may be voided. Contact the manufacturer if in doubt.

Particularly important to any building manager is an accurately engineered drawing of site utilities. The manager can enlist the help of utility companies and plumbing contractors to develop the drawing. (These features are rarely shown to exact finished location on the original construction document.)

The contemporary computer-aided design and drafting (CADD) systems can be very useful to the PM manager. The computer can store information on building systems and update it as needed. It is not necessary to invest in an expensive plotter. The plotting or reproduction can be done by a consultant or a printing service. Both hardware and software systems for computers are so refined now that only the human learning curve and the initial outlay limit the use of automated systems.

After the systems are identified, the next step is to subdivide them into components or workable areas. For the building's roof system, for example, the overall roof area must be subdivided into workable sections that are defined by expansion joints, roof edges, parapet walls, raised flashing, walls, membrane types, deck types, slopes, insulation types, and occupancies. Any difference in the construction, design, or use of a roof area makes it distinctive in terms of life expectancy and PM needs. Often, a professional must be hired to assemble the information to successfully describe a system, assess its condition, and predict its remaining useful life. To obtain the data, the consultant may require nondestructive moisture testing and core sampling.

2.4.4   Developing a Reference Center

Many sources are available to assist the PM staff on how the building should be maintained. These include manufacturers' literature, trade association publications, government documents, your peers, consultants, and educational courses. Trade associations are generally good sources of unbiased information concerning all building and grounds systems. They generally offer conservative non-state-of-the-art recommendations on PM.

2.4.5   Inspection Procedures

Since there are many different building systems, many different individuals are involved in inspection. Inspectors can include maintenance engineers, qualified maintenance craftsmen, contractors, vendors, and consultants. Needs should dictate which inspectors to select.

Inspector selection depends on the technical requirement, the background required, and the budgeted salary. Be careful when hiring an inspector who has a vested interest in the cost of maintenance. In roofing, for example, players have their own special interest. In the case of roof consulting, roofing products manufacturers have a definite conflict of interest when they render opinions to a client. As a group, they are likely to give a biased solution to roofing problems and are likely to serve their own interests in selling their product.

Although roofing contractors offer a variety of systems, from a number of manufacturers, they have biases toward their licensed systems, those they are most comfortable installing. Some contractors have marketing arrangements with manufacturers. For contractors' inspection agents, a great profit potential exists in whatever opinion they give. The larger the potential profit, the larger the potential conflict with the owner's interests.

Testing services offer consulting as an adjunct to their nondestructive moisture testing. Although many of these companies are qualified to find out if moisture is in the system, they may not be qualified to tell whether moisture is system-threatening.

Consultants, engineers, and architects (if they are not representing manufacturers or contractors) represent an uncompromised and objective opinion on system problems. They may, however, have biases toward systems they have designed or installed. They usually tend toward a conservative solution to maximize the probability of success.

Finally, the least compromised inspector is the owner's in-house employee. The only potential drawback is the employee's lack of experience. Unless the in-house inspector is managing a wide variety of systems, he or she may be qualified to deal only with the most obvious system problems.

While your company's inspection instructions concern the in-house staff, the outside consultants must be capable of following inspection procedures established for the in-house personnel. Instructions to inspectors and the PM staff should always be specific and clear. For example, quantifiable terms are needed to explain "tight," the "right gap," and "ponding." To prepare and convey detailed instructions to inspectors and PM personnel is time-consuming, but once you establish clear instructions they can be of substantial service and require only little additional time.

Generally, the PM manager will have some knowledge of how frequently to inspect different building systems. If assistance is needed, various manufacturers' literature, guidebooks, and seminars provide the guidelines adequate for starting inspections. The inspection schedule should be adjusted as needed. With many building envelope systems, inspection frequency is determined by prescheduling the inspection and by external influences, rather than by responses to failures. For example, masonry wall systems are generally inspected once a year. Some building systems must be inspected more frequently or according to insurance requirements. Consider the following frequencies as a starting point:

System	Frequency of Inspection
Roof	Biannual, spring and fall
Grounds, landscaping	Biannual
Walls, foundations	Annual, spring
Walls, sheet metal systems	Biannual, spring and fall
Structure	Annual, spring
Fire alarm systems	Annual
Fire safety systems	Annual
Spectator seating	Annual
Site, mechanical and utilities	Annual
Site, electrical	Annual

Change the inspection schedule according to changing priorities or in response to developing deterioration.

2.4.6 Keeping the Paperwork Simple

A PM program's paperwork is made up of inspection checklists, work orders, equipment logs, and schedule devices. Many times, this area is the cause of PM program failure. Too many forms or ill-conceived forms make procedures unnecessarily complex and overwhelm the staff. Forms must be simple and flexible. Inspection checklists are needed for each building system inspected. These are often available from manufacturers, educational institutes, and trade associations. For roofing, the Roofing Industry Education Institute (RIEI) and the National Roofing Contractors' Association (NRCA) both offer comprehensive checklists (see Figs. 3-12 and 3-13). For masonry inspection, State of Wisconsin, Division of Facilities Management uses a good inspection checklist (see Fig. 8-17).

The key document in the PM program is the work order (Fig. 2-2). It is uniquely numbered, costed, and tracked through the PM program. Like the other forms, it must be flexible enough to be used for different systems. Forms must be kept simple and must be easy to use by all personnel.

2.4.7 Developing an Information Storage and Retrieval System

Information from inspections, work orders, service calls, emergency repairs, and routine repetitive maintenance must be stored in an accessible format.

The format may be a simple flat file where everything is in a folder or it could be a more sophisticated computer file. For smaller projects, manual systems can work effectively. Computer files can be time-consuming to maintain. Generally, they are capable of retaining more data than needed, which can lead to information burnout for users, unless the type and amount of retained data is well thought out. For instance, keep lighting performance on banks of fixtures rather than on individual rooms.

A scheduling system is also needed for the PM activities. It can be a simple manual punch card sorting system, a magnetic status board, or a computerized system. An impressive amount of contemporary software is available for administering and scheduling PM of building systems. Many software programs have integrated work order and reporting documentation.

2.4.8   Establishing Check Points to Monitor Preventive Maintenance Progress

When PM objectives are accomplished they must be documented. Administrative support for a PM program may depend on its documentation. There is no informal system that really works. A formal documentation system tailored to the facility's needs is required.

Periodically review the PM objectives to see if they are being met. Account for the number of work orders and emergency calls for repairs. Establish check points to monitor PM progress by reviewing the amount of repeated work, which may indicate the need for adjusting inspection, PM, and repair schedules.

The PM manager must continuously supervise all PM work and aggressively verify in the field that the work reported has been completed. He or she should carefully investigate any building systems or components showing recurring and frequent PM activities so reasons for this intensive work can be established. For example, the roof drain is always reported as being serviced, yet it continuously has an impressive accumulation of debris, restricting the flow of water. Continuous servicing of this roof drain does not produce desirable results. The cause of debris should then be established and eliminated to ensure that debris does not accumulate.

2.4.9   Performing Corrective Repairs

The building systems included in an established ongoing PM program are assumed to be in a good state of repair. Generally, minor repairs fall into a PM category, but major repairs are not included in the PM activities. PM is not effective on the system needing major repairs or replacement. Systems need to be repaired before PM is initiated or allowed to continue. This hurdle is often the largest in the entire program. Corrective repairs to unmaintained building systems generate large one-time investments that could cost twice the amount of the initial installation. Budget repairs separately from the PM program. Repair work can be easily reported and scheduled on computer. After the necessary major repairs are completed, PM should again be commenced.

2.4.10   Providing Ready Access to the Systems

The need for a safe and convenient access for inspection and maintenance of building systems is critical. A hard-to-access roof, for example, would not be routinely inspected or adequately maintained. Remove wood ladders and replace them with permanently mounted steel access ladders. If uncontrolled pedestrian access to the roof is a problem (and roof access ladders are not to be used for pedestrian traffic), provide a safe extension ladder for staff to use. Do not leave the ladder on the roof to deteriorate.

2.4.11   Managing Changes to the Physical Plant

Changes of physical facilities are unavoidable. These changes could material- ize as additions or modifications of the existing structure. Consider the PM program during the planning of changes. The physical plant director and the PM manager must aggressively review capital improvement projects during planning stages, since those decisions made early in project development have the most profound effect on lowering maintenance cost. Here, the PM mana- ger must play a vital devil's advocate role. If low initial cost systems are proposed that indicate high maintenance costs, a life-cycle cost analysis should demonstrate viability. It will forestall shortsighted design decisions.

2.5   HOW MUCH TO INVEST IN A PREVENTIVE MAINTENANCE PROGRAM

Most building systems last many years before they need to be upgraded, replaced, or dismantled. The cost of a system includes not only the out-of- pocket costs expended at the time of the system's installation but also the costs due to interest on borrowed money, inflation, taxes, and lost revenue (opportunity costs) that result from the system's failure (work stoppage, for example). Money spent for repair and replacement of a system's components, rather than invested for a reasonable return, can also be looked at as lost revenue and should be included when calculating the cost of the system.

The interrelationship of the costs of failure and PM investment and the extent of PM activity is illustrated in Figure 2-5. The figure shows that the more intensive the PM of a system, the less costly the correction of problems (major repairs and replacement due to system's failures) until the cost of PM exceeds the costs of repair and replacement caused by failures. The following list points out the areas of expenditure of funds due to system's failures and PM activities.

Cost of Failure

Contracted roof repairs
Damaged finished goods
Lost rents
Lost profits from residuals
Premiums paid when forced to bid
　in disadvantageous market
　conditions
Increased insurance premiums
Insurance claim deductibles
Lost sales revenues
Damage to structural systems
Damage to finish systems
Attorney fees from suits
Consultant/professional fees
Cost of early roof replacement
Lost interest on invested funds

PM Activities Generating Costs

Man-hours in PM inspections
Annual contracted PM repairs
Periodic nondestructive moisture
　testing (every 5 years)
Consultant inspections
　(every 2 years)
Planned partial flashing or sheet
　metal replacement
　(say, at year 15)
PM inventories of repair materials
Vibration or rotation analyses
Computer hardware and software
　staffing increases

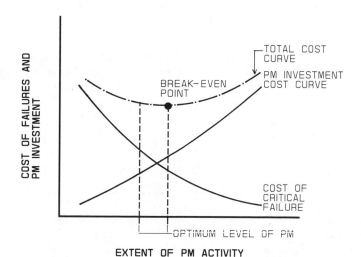

Figure 2-5. Determining the preventive maintenance break-even point.

Figure 2-6 illustrates a practical approach for establishing a break-even point between PM costs and costs of repairs and replacement of the system and the savings that can be realized if the system is maintained. The example (Fig. 2-6) uses a typical computer spreadsheet program to analyze different cost scenarios for maintaining an existing 10-year-old roof system. The following five-step procedure is used for analyzing PM needs:

1. Make assumptions on the life of the system and when costs might occur. Make assumptions on inflation, the cost of money to the firm, and tax rates.
2. List the cash flows in a timetable form. Calculate the inflation and tax effects.
3. Calculate the present value of each cash flow item by multiplying the present value (PV) factor by the item. These factors are commonly found in financial reference tables, as noted in the bibliography. Key the appropriate table with the formula noted in column 8 $(1 + r)$ of Figure 2-6.
4. Add the present values of costs and benefits for each situation analyzed over the life cycle to reach a total net present value. Do this step first for the situation where no PM exists on the roof.
5. Then repeat the process for other sets of strategies (situations), changing the assumptions on expected cash flows and being careful that the net present values of the planned expenditures do not exceed the base situation where there is no PM.

Figure 2-6 assumes that in situation 1 there is no PM on the existing roof. The roof is in critical condition and will fail in 4 years. The other key assumption is that good PM (situations 2 and 3) will push the roof replacement expenditure into year 10. These assumptions are best determined by a roofing professional.

Situation 1 is the base (no PM occurs on the roof). Situation 2 shows the cash flow from a reasonable PM program. Situation 3 shows PM expenditures at the extreme, not-to-exceed limit, where the total net present value equals that of the base ($179,745). These present value numbers (column 9) are not to be taken as absolute expenditure numbers, as noted in column 4. They are valid only for comparisons among different strategies (situations).

The three situations shown in the example illustrate the dramatic effect of pushing back the eventual replacement of the maintained roof an additional 6 years. Significant amounts of money are invested into initial corrective repairs and then that investment is protected with annual expenditures for PM inspections and minor repairs. In situation 3, the PM manager can afford to spend twice the amount of yearly expense than is necessary (situation 2) and will still be as well off as the base situation 1.

Obviously, the PM manager's task is to keep costs to a minimum, and the intent here is not to suggest spending to the break-even point. Nevertheless, this procedure does establish a cost-effective limit.

2.6 WHEN A COMPANY CAN REALIZE PREVENTIVE MAINTENANCE RESULTS

A dramatic, somewhat impractical, way of illustrating the results of a PM program is to compare two building systems: one that has been maintained and one that has not. Since PM minimizes major repairs and replacement of components, and ensures that the system performs to its target date, the comparison should conclusively illustrate that the maintained system outlasts the unmaintained system.

For example, if during a building inspection corroding metal flashing is noticed, an estimate can be made that it will fail within a given time period if not maintained. Painting the flashing will prolong its life to the target date. This procedure is very simple and requires little time. Although direct comparisons of maintained and unmaintained building shell systems are difficult to make because years are required for the benefits of the PM program to become readily apparent, the benefits are being realized.

2.7 HOW TO SELL THE PREVENTIVE MAINTENANCE PROGRAM TO MANAGEMENT

If the benefits of the PM program can be clearly demonstrated to the management, selling the program is easy. To show how successful the PM program can be, the PM manager should demonstrate with data on the building systems that have a history of failure. Typically, some 80% of building problems are caused by 20% of the systems. By choosing a system from the 20% group, the PM manager can clearly show that the failure of the system has been or can be eliminated through a thorough well-managed PM program.

2.8 SUMMARY

Effective PM of buildings requires careful planning, organization, and implementation. The process involves far more than mere reaction by custodial staff to commonly occurring building component failures. To realize PM benefits, substantial effort has to be directed toward organizing an effective PM program and ensuring that different components of the program function efficiently and harmoniously. An efficient program is almost always cost-effective. To realize this benefit, the management of the facility must believe in

Figure 2-6. Example of determining the break-even point for preventive maintenance on a roof system.

(1) YEAR n	(2) EVENT OR COST ITEM	(3) BEFORE TAX COST OF (2)	(4) COLUMN (3) INFLATED 6% OVER TIME	(5) TAX DEPRECIATION (ST. LINE FOR 30 YRS)	(6) TAX BENEFIT (5) X 26% CORP RATE	(7) AFTER TAX INVEST COST (4) - (6)	(8) PRESENT VALUE FACTOR r = 10% n see (1)	(9) PRESENT VALUE (7) X (8)
SITUATION NO. 1 -- LIFE CYCLE ROOF COSTS WITHOUT PM PROGRAM								
1	no work	0	0	0	0	0	1.0000	0
2	emergency repairs	1,000	1,060	0	0	1,060	0.9091	964
3	insurance loss (deductible)	5,000	5,618	0	0	5,618	0.8246	4,633
3	emergency repairs	10,000	11,910	0	0	11,910	0.7513	8,948
4	roof replacement	200,000	252,500	4,208	1,094	251,406	0.6830	171,710
5	no work	0	0	8,417	2,188	(2,188)	0.6209	(1,359)
6	no work	0	0	8,417	2,188	(2,188)	0.5645	(1,235)
7	no work	0	0	8,417	2,188	(2,188)	0.5132	(1,123)
8	no work	0	0	8,417	2,188	(2,188)	0.4665	(1,021)
9	no work	0	0	8,417	2,188	(2,188)	0.4241	(928)
10	no work	0	0	8,417	2,188	(2,188)	0.3855	(844)
TOTALS						NET PRESENT VALUE		179,745

SITUATION NO. 2 -- LIFE CYCLE ROOF COSTS WITH A PM PROGRAM

							NET PRESENT VALUE
1 corrective repairs	10,000	10,000	0	0	10,000	1.00000	10,000
2 PM inspection and repairs	1,000	1,060	0	0	1,060	0.9091	964
3 PM inspection and repairs	1,000	1,124	0	0	1,124	0.8246	927
4 PM inspection and repairs	1,000	1,191	0	0	1,191	0.7513	895
5 PM inspection and repairs	2,000	2,525	0	0	2,525	0.6830	1,725
5 non-destructive moisture survey	3,000	4,015	0	0	4,015	0.6209	2,493
6 PM inspection and repairs	1,000	1,419	0	0	1,419	0.5645	801
7 PM inspection and repairs	500	752	0	0	752	0.5132	386
8 PM inspection and repairs	500	797	0	0	797	0.4665	372
9 emergency repairs	2,000	3,188	0	0	3,188	0.4665	1,487
9 insurance loss (deductible)	5,000	8,448	0	0	8,448	0.4241	3,583
10 roof replacement	200,000	358,160		1,552	356,608	0.3855	137,472
TOTALS			5,969				161,102

SITUATION NO. 3 -- LIFE CYCLE ROOF COSTS WITH A PM PROGRAM (MAXIMUM ADVISABLE INVESTMENT)

							NET PRESENT VALUE
1 corrective repairs	19,094	19,094	0	0	19,094	1.00000	19,094
2 PM inspection and repairs	3,000	3,180	0	0	3,180	0.9091	2,891
3 PM inspection and repairs	3,000	3,371	0	0	3,371	0.8246	2,780
4 PM inspection and repairs	3,000	3,573	0	0	3,573	0.7513	2,684
5 PM inspection and repairs	3,000	3,788	0	0	3,788	0.6830	2,587
5 non-destructive moisture survey	3,000	4,015	0	0	4,015	0.6209	2,493
6 PM inspection and repairs	3,000	4,256	0	0	4,256	0.5645	2,402
7 PM inspection and repairs	1,500	2,255	0	0	2,255	0.5132	1,157
8 PM inspection and repairs	1,500	2,391	0	0	2,391	0.4665	1,115
9 emergency repairs	2,000	3,188	0	0	3,188	0.4665	1,487
9 insurance loss (deductible)	5,000	8,448	0	0	8,448	0.4241	3,583
10 roof replacement	200,000	358,160		1,552	356,608	0.3855	137,472
TOTALS			5,969				179,745

PM and must be willing to allocate resources for it: The management must know how much to invest in the program. This step is the foundation of an effective PM program.

Next the management should outline the objectives of the proposed program. What should the program encompass? What facility needs should it address? After these steps, the program itself should be structured. This stage should take into account such concerns as qualifications and training of the PM manager/supervisor and staff, implementing information storage and retrieval systems and a reference center, prioritizing the PM needs of the facility, establishing effective inspection procedures, and specifying proven PM and repair practices to follow. Since benefits are not always readily apparent, the management of the facility must be kept informed of the potential problems that were averted by PM. This technique helps to ensure program support.

Most people need some convincing about the importance of preventive measures whether for building maintenance or for health maintenance. Humans tend to respond slowly to preventive measures. Yet, innumerable PM programs have proven that maintaining a building is cost-effective. It protects the owner's investment and prevents unexpected failures of building systems that disrupt facility activities. An effective PM program is not easy to structure and implement, but once in place it is an asset to any organization.

2.9   SUGGESTED READINGS

Bierman, Harold, Jr., and Seymour Smidt. 1971. *The Capital Budgeting Decision*, 3rd ed. New York: Macmillan.

Criswell, John W. 1989. *Planned Maintenance for Productivity and Energy Conservation*, 3rd ed. Lilburn, GA: Fairmount Press.

"Maintenance Library." *Plant Engineering*. 1987. A collection of reprint articles. Des Plaines, IL: Cahners Publishing Company.

The New Good School Maintenance, a Manual of Programs and Procedures for Buildings, Grounds, Equipment. 1984. Springfield, IL: Illinois Association of School Boards (IASB).

School Facilities Maintenance and Operations Manual. 1988. Reston, VA: Association of School Business Officials International.

White, John A., Marvin H. Agee, and Kenneth E. Case. 1984. *Principals of Engineering Economic Analysis*, 2nd ed. New York: Wiley.

2.10   ASSOCIATIONS WITH INFORMATION ON ORGANIZING PREVENTIVE MAINTENANCE ACTIVITIES

American Institute of Plant Engineers
3975 Erie Avenue
Cincinnati, OH 45208

American Society for Hospital Engineers
American Hospital Association
840 North Lake Shore Drive
Chicago, IL 60611

Association of Physical Plant Administrators of Universities and Colleges
1446 Duke Street
Alexandria, VA 22314–3492

Association of School Business Officials (ASBO) International
11401 North Shore Drive
Reston, VA 22090–4232

International Maintenance Institute
P.O. Box 266695
Houston, TX 77207

3. Built-Up Roofs

Paul Tente

3.1 INTRODUCTION

Bituminous roofs had their beginning in the mid-nineteenth century. The first built-up roofs (BUR) were made of paper and pine tar. Today this type of roof is made of felts and hot bitumen, which can be either coal tar pitch or asphalt. The bitumen or adhesive provides necessary weather proofing and the felts provide reinforcement. To obtain a built-up roofing system, layers of felts are mopped together with bitumen, creating a highly waterproof and durable membrane (Fig. 3-1).

Before World War II, construction of built-up roofs was governed primarily by field experience. Very little scientific research took place. In recent years, however, much scientific work on roofing has been done by numerous organizations including the National Bureau of Standards (NBS) and the National Roofing Contractors' Association (NRCA).

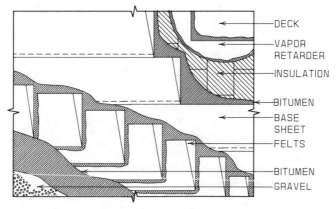

Figure 3-1. Built-up roofing construction.

From the moment of installation, the roofing system undergoes continuous deterioration. Extreme temperature fluctuations as well as snow, ice, hail, and wind prevail upon the roofing surface. Traffic on the roof and the installation of mechanical and other types of equipment can cause physical damage that contributes significantly to roofing failure.

A well-functioning and durable roof depends on good design, appropriate and correctly installed materials, and maintenance. A roof is manufactured in place, which in many instances is done under poor environmental conditions with workers who do not have adequate experience. Conditions for potential problems are always present during new construction and maintenance. To ensure a properly performing roof, roofing designers, installers, and repair personnel must recognize these limitations and know how to deal with them.

Of the requirements (design, materials selection and installation, and maintenance) for a well-performing roof, preventive maintenance (PM) of the roofing system generally receives the least amount of attention and yet it can extend the life of the roof for many years. PM does not mean major repairs or replacements of components of the roofing system. It is performed to prevent possible problems by regularly inspecting the system and maintaining it. For a PM program to be effective, it must be well organized, possess continuity, employ qualified staff, and be adequately budgeted. Any effective PM program for BURs should include the following:

Record keeping: Drawings and specifications of new roof construction, as well as any repair and maintenance work done at various periods, should be available.

Qualified staff: A company should either maintain qualified in-house staff for doing PM roofing work or hire reputable, qualified, outside contractors.

Inspection: Any effective PM program should include regular inspections of the roofing system.

Design, materials, and methods of application: The staff responsible for PM should understand and be skilled in the design of maintenance work and selection of maintenance materials and their application.

Budgeting: Adequate PM work cannot be performed if the budget is inadequate. The dollars spent on PM of the facility roof can be a wise investment. Although appropriate materials and skilled workmanship during new construction can contribute significantly to avoiding premature failures, it is through an effective PM program that premature failures can be eliminated.

3.2 COMPONENTS OF BUILT-UP ROOFING SYSTEMS*

To minimize the need for excessive and costly PM of BURs, it is imperative that roofing systems are appropriately designed and that the materials are installed according to the recommended roofing industry standards. In BUR, as in all building systems, the quality of original construction dictates largely PM and replacement requirements. Continuous maintenance of inappropriately designed and installed elements is counterproductive. Base flashings that are too low or drains that are located on the high points of the roof are examples of faulty design and installation that may be responsible for excessive maintenance and unending future problems. Usually, PM procedures are similar to the procedures of new construction. Understanding new construction, therefore, allows for an effective performance of PM work.

Conventional hot-applied BUR is comprised of four basic components: bitumens, felts, surfacings, and flashings. Three components necessary to the BUR system, although not directly related to the waterproofing of buildings, include substrates (structural roof decks), vapor retarders, and insulation. Because these three components are not used for waterproofing of buildings, they are most often thought to be parts of other building systems and not parts of the BUR system. Roof bonds, for example, specifically exclude these items. In the past, the approach to proper specification and installation of these items has generally been haphazard at best. There is not a clear understanding by architects, specifiers, contractors, and sometimes even roofing manufacturers of the direct effect that each component has on proper roofing performance. For these reasons, and especially because building insulation has become increasingly important in energy conservation efforts, it is imperative that these underlying components of the BUR system be reviewed before dealing with roofing maintenance concerns.

3.2.1 Roof Decks

The design of a roofing system must always take into consideration the substrate. The various conditions of anchoring and supporting the components of the roofing system start here. The substrate should be designed to provide positive drainage. It should not deflect excessively and cause ponding. The substrate should be well anchored, structurally sound, and dimensionally stable to preclude excessive movement injurious to the balance of the

---

*Section 3.2 was contributed by Arthur "Pete" Simmons.

system components. In many cases, expansion joints must be provided in the substrate to prevent failures at the following points: where rebuilding changes direction (as in L-shaped, U-shaped, and H-shaped buildings); where framing changes direction; where there is a change in substrate material (e.g., from steel to concrete); where there are additions to an existing building; and where there are changes of elevation between adjoining decks.

Before installing the roofing membrane, it should be ensured that the substrate is dry or that adequate ventilation is provided for dissipation of any moisture. The roof deck must be uniformly level. Substrates with ridges, depressions, or high–low levels between adjacent units should not be accepted. The substrate should be conducive to positive attachment of the roofing components, such as flashings, and gravel stops. The installation of all openings, projections, and drainage systems on the roof should be completed prior to the installation of the roofing membrane. Inspection and acceptance of the substrate should be made by a qualified inspector or testing service and generally not by the roofing subcontractor.

3.2.2   Vapor Retarders

It would seem that this component (from its name alone) should adequately describe its purpose, use, and even its installation procedures; however, this is not always the case. Where to install a vapor retarder, how and when to install it, and what alternatives exist for the use of vapor retarders are questions that in many instances remain unanswered.

In general, the roofing industry, both in manufacturers' specifications and on-site installation, ignores the fact that for the vapor retarder to be effective, it must be continuous. Since roofing manufacturers' specifications do not require or even mention sealing of the vapor retarder at drains, pipes, and/or other projections, and since roofers often do not understand the purpose of the vapor retarder, the result is that a continuous vapor retarder is seldom, if ever, installed. If a vapor retarder is not continuous, its effectiveness is negated and it can be damaging to the roofing system.

Water vapor is present at all times in our atmosphere. It is colorless, dry gas. When the vapor from the inside of an enclosed building migrates toward the outside (lower temperature) it reaches a zone (a dew point) where it condenses. On an insulated roof, the dew point is normally somewhere within the roof insulation or at the underside of the roof membrane. If water vapor is allowed to pass through penetrations in the vapor retarder, the moisture could spread and condense within the insulation, causing serious problems: The efficiency of the insulation can be greatly reduced; the freezing of the moisture at the underside of the roofing membrane can release the membrane from positive bond to the insulation. Thus, blistering, buckling, and similar defects can occur.

Do not specify that "the vapor retarder shall meet Factory Mutual Class One construction requirements," or "the NRCA Manual Guidelines" and leave it at that. Specify that "the vapor retarder should be positively sealed at side and end laps at all projections, drains, and all intersections of vertical planes and the roof deck." Since a vapor retarder having a zero perm rating is difficult to obtain under normal construction practices, the prudent designer will vent the roof where appropriate in addition to providing a vapor retarder.

A roofing system that allows for subroof ventilation is most effective for preventing accumulation of moisture in BUR insulation. However, the installation of a vapor retarder and a venting system can add significantly to the construction cost and, in some cases, it is not necessary. Normally a vapor-retarder should be used when the January mean temperature is 35° F or lower and when the occupancy of a building results in relative humidity above 40%. A low-humidity occupancy (under 20% RH) is usually found in dry storage warehouses. Generally, a medium-humidity occupancy (20–45% RH) exists in such building types as auditoriums or apartments, and high-humidity occupancy (over 45% RH) can be found in such building types as breweries, food processing plants, and paper or textile mills.

3.2.3 Roof Insulation

Roof insulation, regardless of type, must be positively secured to the roofing substrate. If proper securement of insulation is not achieved, two types of failures can occur: potential wind damage or blowing off of the roofing ; or splitting of the BUR membrane. The splitting of the membrane is due to unrestrained thermal expansion of insulation, or to cupping or raising of insulation board edges.

Proper attachment of insulation can be accomplished with hot bitumen or mechanical fasteners. Factory Mutual (FM) System requirements for steel decks call for mechanical fasteners. As with vapor retarders, field application practices of roof insulation are often inadequate, again probably because insulation is not considered to be a part of the waterproofing system. Many premature roof failures are directly attributable to insulation installation errors.

Two methods for insulating a roof are in use. One is placing the insulation over the roof deck. The other method is insulating below the deck. Given the choice for a structure adaptable to either type of insulation installation—rigid board above deck or batt/blanket below deck—the batt/blanket system is usually preferable. Below-deck insulation is considerably less expensive. Also rigid board above-deck insulation has some deleterious effects on BUR performance. These undesirable effects include the increased chance for con-

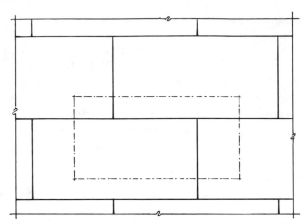

Figure 3-2. Rigid insulation should be installed with staggered joints.

densation (or entrapped moisture) within the roofing system, the acceleration of bitumen hardening and oxidizing from high roof temperatures during summer weather, and the production of greater thermal changes in the roof membrane (superheating and subcooling). The thermal changes create stresses on the membrane's system and increase the possibility of the roof splitting due to temperature extremes between summer and winter.

With certain types of roof decks such as steel, rigid board over-the-deck insulation is a necessity. Also, on other types of decks such as wood fiber, the use of rigid board over-the-deck insulation may be advisable to act as a buffer between the roofing membrane and the joints of deck panels. It is recommended that rigid roof insulation be installed in two layers, with the joints staggered between the courses (Fig. 3-2).

Where Urethane, Polyisocyanurate, and Phenolic Foam roof insulations are used, a top layer of insulation should consist of materials such as Fiberboard, Fiberglas, or Perlite if hot-applied BUR is to be installed over the insulation. As in all two-layer insulation systems, the joints of the top-layer insulation system should be staggered from joints of the bottom layer. The top layer of roof insulation in a two-layer system is generally cemented to the bottom layer using hot asphalt.

Rigid roof insulation should possess the following properties:

1. Compressive strength sufficient to withstand normal rooftop traffic.
2. Dimensional stability against thermal and moisture changes.
3. Cohesive strength to resist delamination resulting from wind uplift.
4. Horizontal sheer strength to resist tensile stresses of the roof membrane.
5. Ability to accept and hold adhesives both to the roof deck and to the roof membrane. The principal types of roof insulation with some of their material properties are shown in Table 3-1.

Table 3-1. Properties of Roof Insulation Materials

Material	Permeance (perm, in.)	Density (lb/cu ft)	k Factor	R Factor	Coefficient of Thermal Expansion in/in./ $°F \times 10^{-6}$	Comprehensive Strength	Fire Resistance
Wood (cane) fiberboard (organic)	20–50	15–22	0.36	2.78	3	Good	Poor
Cellular glass (inorganic)	022	22	0.35	2.86	4.6	Excellent	Excellent
Fiberglas board (inorganic)	high	9	0.26	3.85	—	Poor	Good
Foam polystyrene (organic)	1.2–5.8	1.9	0.26	3.85	35	Fair	Poor
Perlite aggregate board (inorganic)	25	11	0.38	2.63	Negligible	Good	Excellent
Foam urethane (organic)	11.5–22.5	1.5–2.5	0.17	5.88	30	Fair	Poor
Polyisocyanurate	11.5–22.5	1.5–2.5	0.17	5.88	30	Fair	Good
Phenolic Foam	7.0	2.5	0.12	8.3	13	Fair	Good

3.2.4   Bitumen

The bitumen in a conventional BUR system is the waterproofing agent. Asphalts and coal tar pitch are two primary classes of bitumens used. Asphalt is a natural constituent of most crude oils. To obtain asphalt, crude oil is heated and distilled so volatile products such as gasoline, naphtha, kerosene, and heating oils are removed. The asphalt is air-blown to raise softening points to four types of bitumen suitable for roofing construction:

1. Type I, dead-level asphalt (also flat, level-deck, Aquadam, etc.), 135–150° F, slope ranges: up to ½ in./ft
2. Type II, low melt, 160–175° F, slope ranges: ½ to 1½ in./ft
3. Type III, high melt (steep), 180–200° F, slope ranges: ½–3 in./ft
4. Type IV, V., high melt (special steep), 205–225° F, slope ranges: ½–6 in./ft

Coal tar pitch is obtained from coal by destructive distillation of coal and the fractional distillation of coal tars. Coal tar pitch is not air-blown as asphalt is and it is delivered with only one range of softening points: 140–155° F with a normal safe slope range limited to 0 to ½ in./ft.

Both types of bitumens have several properties that contribute to making them practical for use in roofing construction. Both prevent passage of water vapor, except minute amounts, and do not conduct surface water. Recent tests show that after 2 years of immersion in water, asphalt absorbed approximately .1 gallon of water per 100 ft² per year and coal tar pitch absorbed only one half that amount. Both bitumens lose absorbed water rapidly when surface-dry.

The bitumens are thermoplastic in nature. They can be applied by simple heating. After application, they harden to perform their adhesive and waterproofing functions. Bitumens are soluble in common solvents, such as mineral spirits, and can be used in a variety of cold-applied coatings and adhesives. These coatings and adhesives set by evaporation of solvents. Bitumens are also dispersable in water and, in combination with mineral colloids, form coatings and adhesives that set by evaporation of water. Bitumens adhere well to roofing felts of all types and to almost any dry surface. They bind the roofing membranes together and are used for adhering roofing insulation and vapor barriers to decks.

Controlled temperatures during application of bitumens are necessary to retain waterproofing and adhesive characteristics. If overheated, bitumens can lose volatiles, become hardened, and lose both adhesive and waterproofing properties. Overheated asphalts suffer in two ways: Initial overheating may cause some softening point, which may result in roof slippage. Prolonged overheating may "crack" the product, leaving a residue like coke. Asphalt when overheated becomes progressively more fluid, easier to mop, and it spreads more thinly, thus tempting the applicators to overheat the material. Recent tests indicate that asphalt may be heated above normal top limits for a limited time without adverse effects. In practice, however, it would require a full-time inspector at the kettle with a stop watch to make certain that asphalt is not overheated for a longer period of time than acceptable. For this reason, it is recommended that maximum temperatures are adhered to. Follow equiviscous temperature (EVT) guidelines. Overheating does not increase the fluidity of coal tar pitch.

Although there is little coal tar pitch bitumen used for roofing today, pitch is still used extensively for maintaining existing pitch roofs. Therefore, it is important to understand the difference between asphalt and coal tar pitch. Coal tar pitch has excellent coal-flow self-healing properties. The pitch flows together and heals cracks. Pitch is self-priming whereas asphalt is not. However, because of the cold-flow characteristics and only the one softening point grade, the use of coal tar pitch is limited to low-sloped applications, usually not over ½ in./ft. Because of the lower water absorption and self-healing properties, coal tar pitch built-up membranes perform better on flat roofs where ponding and more continual standing water occur. Coal tar pitch, however, is more susceptible to oxidation and hardening under direct exposure to the sunlight than asphalt. Therefore, coal tar pitch must be protected and should be surfaced with aggregate, gravel, or slag.

Asphalt, because it is available in a number of grades (softening points), is more adaptable to different roof slopes. It does not self-prime, however, and its self-healing characteristics are not as good as those of coal tar pitch. Asphalt roofing manufacturers do not recommend its use in areas of ponding water. Asphalt weathers better than pitch, oxidizing more slowly upon exposure to the sun's rays. Hot asphalt or a variety of asphaltic coatings can be used in place of gravel for surfacing.

Asphalt and coal tar pitch are not compatible; they cannot be used together. The two bitumens should never be mixed nor should pitch felts with asphalt moppings or asphalt-saturated felts with pitch moppings be specified. The exceptions are that asphaltic base flashings can be used with coal tar pitch roof membranes, and asphalt-coated base felts—stabilized asphalt coatings and minerals—may be used with pitch moppings without bad effects. All surfaces must be dry to receive bitumens.

3.2.5 Felts

The conventional BUR membrane is made of alternate layers of felts and bitumen. Roofing felts are comprised of two basic types of material: organic fibers and glass fibers. Asbestos fiber felts are no longer used.

Organic fibers are manufactured from felted papers and wood pulp, then saturated with either asphalt or coal tar pitch. In some areas of the United States, they are still known as "rag felts" from the past practice of using rags in addition to cellulose fibers. A dry felt weighing approximately 5.5 lb/100 ft^2 becomes the "15-lb" felt when saturated. Glass fiber felts are made from a loosely woven mat of threadlike inorganic glass fibers held together with a binder. Typical dry glass felts weigh about 1½ lb/100 ft^2 and are combined with asphalt to make a roofing felt of about 8 lb in weight.

From the basic roofing felts, which vary in weights from approximately 8 lb to 30 lb, two other products are made for use in the roofing industry: coated felts and mineral-surfaced cap sheets. Coated felts are made by factory-applied asphalt coating on both sides of the saturated felts with a releasing agent such as talc, silica, sand, or mica added to either one or both sides (to prevent sticking in the roll). Mineral-surfaced cap sheets are made by adding a hot coating and embedding mineral granules on one side of the felt. Granule colors are available in a wide range. The process used to manufacture organic felts organizes lengths of fibers into a longitudinal direction rather than across the roll. This case is also true, to a somewhat lesser degree, with glass fiber felts. This factor must be considered when placing felts over substrates. In the case of each felt type, but particularly with regard to organic felt, strength is greater in the longitudinal direction than in the transverse direction.

Usually, multiple layers of felts are used for roofing systems, permitting the application of the waterproofing agent, bitumen, in individual layers and adding strength and overall flexibility to the system. The multiple layers also provide some added mechanical damage resistance to the roofing membrane, distributing the stresses over a wider area. They provide a margin of safety for skips or omissions in any of the several moppings.

Moisture is the common enemy of all BUR systems. Any felts over which surfacing has not been completed on the day of their installation should be protected with a glaze coat of asphalt at the end of the day.

3.2.6   Surfacings

There are three basic types of roof surfacings: mineral aggregates, hot asphalt or cold-applied coatings, and mineral-surfaced cap sheets. Mineral aggregate is commonly gravel or slag, but may include dolomite or other types of rock. The primary requisites of surfacing aggregates are that they must be opaque, clean, and dry. They must be properly sized to allow for their appropriate embedment into the flood coat to provide protection from the sun's rays.

Mineral aggregate is most commonly used on lower slopes, 2½ in./ft being the practical maximum. Normally, all gravel- or slag-surfaced roofs over any type of substrate qualify for U/L (Underwriters Laboratories) class A fire rating.

The roof can also be smooth-surfaced. For such a roof three types of coatings are available: hot asphalt, applied at 20–25 lb/ft^2; asphalt clay-type emulsions, applied 3–6 gal/ft^2; and cut-back (solvent blends), applied at a rate of 1½–3 gal/ft^2. The smooth-surfaced roof should be installed only on slopes where good drainage is provided—¼ in./ft. Fire ratings for smooth surfaces will vary from U/L class A on one extreme to no fire rating with certain organic felt membranes on the other. Smooth roofs require more frequent maintenance than graveled roofs and have to be recoated every 3 to 6 years to protect the roof membrane.

As with smooth-surfaced roofs, minimum slopes for mineral-surfaced cap sheets should be ¼ in./ft for inorganic cap sheets and ¾ in./ft for roofs with organic cap sheets. Mineral-surfaced cap sheets can be obtained in varied colors for appearance purposes and reflectivity. Fire ratings for cap sheets vary from U/L class C to U/L class A, depending on the membranes and substrates. At lower slopes, cap sheets will normally require maintenance every 5 to 7 years.

Although aggregate-surfaced roofs require less periodic maintenance, have a greater resistance to damage from wind or hail, and perform better under low slope conditions, smooth-surfaced roofs and cap sheets have some advantages. They are lighter in weight—approximately 4 lb/ft—and are more easily repaired. Damage to the smooth-surfaced roofs is not concealed, as is generally the case on the aggregate-surfaced roof.

3.2.7 Flashings

Flashings are responsible for more than 80% of all roof leaks because of faulty design, incorrect specifications, and faulty installation caused by the lack of understanding of the function of flashings. Where the BUR membrane is interrupted or terminated, flashing should be installed (Fig. 3-3). It should be watertight and weathertight, wind resistant, durable, and able to withstand the differential stresses of the dissimilar surfaces it adjoins.

Base flashing is the industry term for composition flashing, which is installed along the cant strips at all intersections between the vertical surface and the roof deck (Figs. 3-4 and 3-5). This flashing usually consists of multiple layers of roofing felts (from two to four), cemented either by hot asphalt or fibrated asphalt flashing compound. The surfacing sheet is mineral-surfaced cap sheet for weathering performance. Base flashing may also be comprised of elastomerics or thermoplastics, usually applied in a single layer and often installed with special adhesives designed for compatibility both with the bituminous roofing and the flashing material.

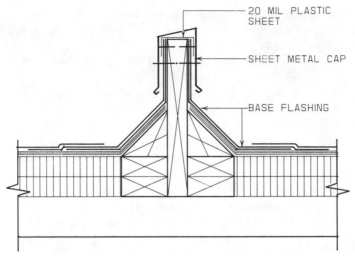

Figure 3-3. Correct control joint installation.

To help ensure good flashing design, follow these recommendations:

1. Provide 8 in. minimum flashing height; 10 in. are recommended in areas of heavy snowfall.
2. On concrete and masonry surfaces (e.g., parapet walls) always provide for a positive seal at the top of base flashing.
3. Provide mechanical fasteners or nailing at the top edge of flashing to prevent slipping or pulling away from vertical surfaces.
4. Make allowances for movement, vibration, and maximum attachment.
5. Specify and design a good system composed of quality materials. It is usually not enough to specify a "20-year bondable system."
6. Always use cant strips with base flashing.
7. The top edges of base flashing should be protected by the use of counter-flashing.

Auxiliary metal flashing includes gravel stops, pipe and conduit flashing, and pitch pans, and should be equipped with minimum 4-in.-wide flanges for incorporation into the roofing system. The flanges should always be set over the top of installed felts, fastened, and then stripped in with additional layers or collars of felt and fabric, prior to application of the roof surfacing. Do not locate flanged auxiliary flashing in areas where ponding water may occur. Gravel stops and other perimeter metal flashings are best installed on raised curbs at roof edges (Figs. 3-6 and 3-7). Pitch pans, because of the necessity for continuing maintenance, should be used to flash roof penetrations only as a

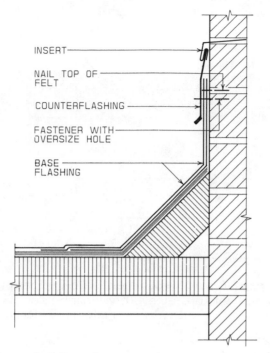

INSERT

NAIL TOP OF FELT

COUNTERFLASHING

FASTENER WITH OVERSIZE HOLE

BASE FLASHING

Figure 3-4. Base flashing installation: brick wall.

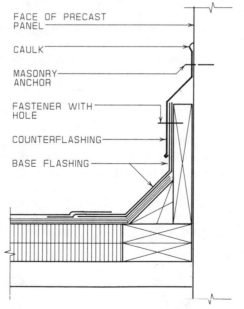

FACE OF PRECAST PANEL

CAULK

MASONRY ANCHOR

FASTENER WITH HOLE

COUNTERFLASHING

BASE FLASHING

Figure 3-5. Base flashing installation: concrete wall.

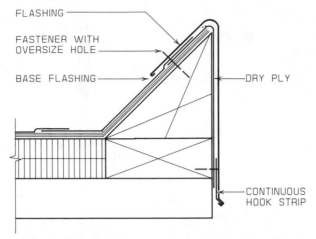

FLASHING

FASTENER WITH
OVERSIZE HOLE

BASE FLASHING

DRY PLY

CONTINUOUS
HOOK STRIP

Figure 3-6. Typical recommended roof edge detail.

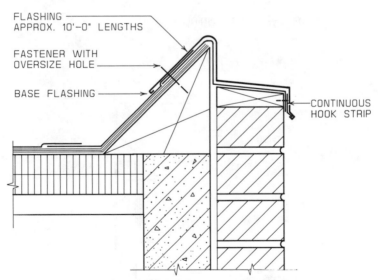

FLASHING
APPROX. 10'-0" LENGTHS

FASTENER WITH
OVERSIZE HOLE

BASE FLASHING

CONTINUOUS
HOOK STRIP

Figure 3-7. Roof edge detail designed to appear as a thin line.

last resort. The better approach is to build properly flashed curbs. Expansion joint covers should be installed on curbs raised above the roof level.

3.2.9   Roofing Installation Recommendations

It is critical to follow good roofing installation practices, whether in new construction or in roof repairs and maintenance. Lay roofing plys shingle-fashion starting at low points. Broom all felt layers full-width to eliminate trapped air and to ensure proper interply lamination. Make immediate repairs to fish mouths, wrinkles, buckles, tears, and so forth in each felt layer as work proceeds. Be certain that interply moppings are uniform and complete, with no skips or voids. Store all job site materials on raised pallets with weather-protective covering. Use equipment with soft pliable tires for moving materials on the roof. Apply roofing only in dry weather and, generally, at temperatures of 40° F or above.

3.3   CAUSES OF BUILT-UP ROOFING FAILURES

In order to obtain a BUR system that performs effectively for the duration of the expected system's life, factors that cause failures must be clearly understood. Some of the roofing problems are built in during new construction and are, generally, out of the hands of PM personnel. Many potential problems, however, can be eliminated by the maintenance staff, providing they understand how the BUR system functions and why it fails. In general, causes of BUR failures can be divided into a number of categories: faulty original design, incorrect installation of components, modifications of existing roofing without knowing roof-loading capabilities or other restricting factors, abuse of the roofing system, and weathering.

Inadequate slope or excessive deflection can cause serious ponding. A lack of the necessary expansion joints can cause splits. Improperly prepared and assembled components, the placement of insulation on uneven decks, or an inadequate amount of adhesive to anchor insulation to the deck could result in water entry into the building. Wet materials or vapor retarders improperly used also make the roof highly susceptible to severe problems. The second and third layer of roofing membrane on top of the original roof and additional mechanical units on the roof may overload the roof. When changes in the mechanical systems on the roof and the modifications of the roofing system are first contemplated, it is imperative that the structural system of the building be checked to ensure the building is capable of accommodating the load.

One of the main causes of roofing failure is roof abuse (Figs. 3-8, 3-9, and 3-10) caused by traffic and work related to the maintenance and repair of

Figure 3-8. Roof abuse. *(Courtesy: Division of State of Wisconsin Facilities Management.)*

parapets, mechanical equipment, piping supports, and so on. This problem is seldom considered during new construction and yet damage to the roof from the foot traffic and other abuse is easily prevented. During new construction appropriate walks and mechanical equipment repair areas should be constructed, periodically inspected, and adequately maintained.

Preventing potential problems from normal weathering is, of course, the main focus of PM. Damage caused by wind, hail, and pollutants should be arrested immediately. Otherwise, problems could develop that require major and expensive repairs.

3.4 INSPECTION

An essential part of a PM program is inspection. In new construction the installation of the roofing system generally requires daily inspections. A qualified inspector should be on the site during the roof installation process. The inspector should pay attention to the type of equipment used by the contractor, the condition of materials, and installation procedures.

After the roof is completed, it must be inspected at least twice a year and after any type of damaging storm. If the roof has mechanical equipment that requires maintenance, these areas should be inspected much more frequently. Such frequency is necessary to ensure that mechanical maintenance crews do not jeopardize the weathertight integrity of the building. The same person should make all inspections. Single responsibility is far better than a

Figure 3-9. Pipe support digging into the roofing.

Figure 3-10. Oil leaking from mechanical equipment, causing rapid degradation of bitumen.

committee-type approach. Group responsibility, in many cases, ends up with no responsibility.

The first step of inspection should be to inquire whether there are any water-entry problems. If water entry is present, find out what conditions cause water to enter the building. Quite often, water will only come in under certain conditions and those conditions will offer a clue about the problem's source. Note that all water entries into a building are not necessarily caused by a leaking roof. Water problems can be caused by condensation, negative air pressure within the building, or entry through walls or mechanical equipment (Fig. 3-11). Problems caused by condensation are common. When inspecting for water entry ask the following questions.

1. Does the leakage occur basically during the winter months?
2. Is the problem above a swimming pool, hot tub, laundry, or kitchen?

These findings may indicate condensation. Check for water entry during periods of heavy rain. If the water problem is not intensified after heavy rains, then it might also be due to condensation.

Once water has entered the roofing system it is imperative to define the extent of damage, especially wet insulation. The simplest method involves the use of an inexpensive instrument, the moisture meter. Since its use requires puncturing of roofing membrane, this method has limitations, but it is a useful approach for verifying suspected wet roofing sections. For a more complete investigation of conditions caused by water entry into the roofing system, test cuts may be necessary (Fig. 2-4). This technique allows you to

Figure 3-11. Water entry into the building can be caused by cracks or failing joints in the walls.

examine in detail various components that have been affected by moisture. But here, too, test cuts are limiting since they require cutting through the roofing assembly. For locating wet areas, these two methods are only marginally helpful.

Nondestructive evaluation of moisture in roofs has been gaining acceptance and is the most reliable approach for detecting general areas of moisture. In the mid-1970s, there were just two or three firms performing nondestructive roof tests; now there are 300 or more. The three nondestructive methods employed for defining wet areas in the roofing system include capacitance metering, nuclear metering, and infrared scanning (see Table 4-1).

When inspecting the roof include a thorough walk of the whole roof area. Check to see if there are any changes such as different ponding patterns or greater deflection that were not present during the previous inspection. Check for drain blockage. Look for possible abuse of the roofing surface from foot traffic, especially around the mechanical equipment units. Look for excessive weathering of the roof surface, observing flashings, gravel stops, and roof penetrations in general. Figures 3-12, 3-13, and 3-14 (on pages 54–62) are examples of BUR inspection checklists. With some modification they can be made applicable to most any situation.

3.5   PERFORMING PREVENTIVE MAINTENANCE

Before starting roofing maintenance, if the roofing system is under a warranty from either contractor, manufacturer, or both, contact them immediately and explain your findings. Have them come out to confirm the conditions and the need for maintenance. Even if normal maintenance is not the responsibility of the manufacturer or contractor, they should be made aware of its need and should give their recommendations. The repairs or maintenance should not affect the remainder of the warranty.

Whenever you perform repairs or maintenance, roofing manufacturers' specifications and details should be followed and adhered to diligently. Keep in mind that PM and repair procedures are similar to the procedures of new construction. If specifications and details for repair and maintenance work are unavailable, follow those recommended for new construction. For this reason, PM personnel should be skilled not only in roofing maintenance but in roofing construction in general.

When performing PM follow these recommendations:

1. The areas worked on and materials used in PM should be clean and dry.
2. It is important to ensure that materials used are compatible; if the roof felts are mopped together with pitch, then pitch (not asphalt) should be used in maintenance or repair work.

(Text continues on page 63.)

Figure 3-12. A built-up roofing inspection checklist recommended by the Roofing Industry Educational Institute. (Reprinted by permission of the Roofing Industry Educational Institute.)

Semi-Annual Maintenance Inspection Checklist

Building _____ Date _____ Date of previous inspection _____

Location _____ Inspected by: _____

I. Supporting Structure	OK	Problem Minor	Major	Observation	Date of Repair
Exterior and Interior Walls					
Expansion/Contraction					
Settlement Cracks					
Deterioration/Spalling					
Moisture Stains/Efflorescence					
Physical Damage					
Other					
Exterior and Interior Roof Deck					
Securement to Supports					
Expansion/Contraction					
Structural Deterioration					
Water Stains/Rusting					
Physical Damage					
Attachment of Felts/Insulation					
New Equipment/Alterations					
Other					

II. Roof Condition

A. General Appearance
Debris
Drainage
Physical Damage
General Condition
New Equipment/Alterations
Other

B. Surface Condition
Bare Spots in Gravel/Ballast Displaced
Alligatoring/Cracking
Slippage
Other

C. Membrane Condition
Blistering
Splitting
Ridging/Wrinkling
Fishmouthing
Loose Felt Laps/Seams
Punctures, Fastener Backout
Securement to Substrate
Membrane Shrinkage
Membrane Slippage
Other

III. Flashing Condition

A. Roof Perimeter Base Flashing
Punctures or Tears
Deterioration:
Blistering
Open Laps
Attachment
Ridging or Wrinkling
Other

Flashing Condition (Continued)

55

Figure 3-12. (Continued)

	OK	Problem Minor	Problem Major	Observation	Date of Repair
III. Flashing Condition (cont'd.)					
B. Counter Flashing/Termination Bars					
Open Laps					
Punctures					
Attachment					
Rusting					
Fasteners					
Caulking					
Other					
C. Coping					
Open Fractures					
Punctures					
Attachment					
Rusting					
Drainage					
Fasteners					
Caulking					
Other					
D. Perimeter Walls					
Mortar Joints					
Spalling					
Movement Cracks					
Other					
IV. Roof Perimeter Edging/Fascia					
Splitting					
Securement					
Rusting					
Felt Deterioration					
Fasteners					
Punctures					
Other					

V. Roof Penetrations

A. Equipment Base Flashing-Curbs
   Open Laps
   Punctures
   Attachment
   Other

B. Equipment Housing
   Counter Flashing
   Open Seams
   Physical Damage
   Caulking
   Drainage
   Other

C. Equipment Operation
   Discharge of Contaminants
   Excessive Traffic Wear
   Other

D. Roof Jacks/Vents/Drains
   Attachment
   Physical Damage
   Vents Operable/Screens Cleaned
   Other

VI. Expansion Joint Covers
   Open Joints
   Punctures/Splits
   Securement
   Rusting
   Fasteners
   Other

VII. Pitch Pockets
   Fill Material Shrinkage
   Attachment
   Other

57

Figure **3-13.** Built-up roofing inspection from *NRCA/ARMA Manual of Roof Maintenance and Repair.* *(Reprinted by permission of National Roofing Contractors Association and Asphalt Roofing Manufacturers Association.)*

INSPECTION FORM—BUILT-UP ROOF   No.

BUILDING ID: _____   INSPECTION DATE: _____

HISTORICAL RECORD NUMBER: _____

ROOF DECK UNDERSIDE:
Record locations of signs of water leakage/damage: _____
_____
_____

Record locations and extent of deck deterioration: _____
_____
_____

Record locations of bitumen drippage: _____

PARAPET/WALL EXTERIOR:
Deteriorated Mortar Joints _____   Settlement Cracks _____
Open Coping Joints _____   Cracked/Broken Coping Cap _____
Eflousescence _____   Damaged Facia/Overhang _____
Deteriorated Gutters, etc. _____   Other _____

58

Record locations of signs of water infiltration: _____

_____

_____

FIELD OF ROOF:

General Condition: Good _____ Fair _____ Poor _____

Watertightness:

No leaks _____ Leaks during continued rain _____

Leaks every rain _____ Leaks during high wind _____

Leaks continuously _____

Check for evidence of: (indicate level of damage as L=light, M=moderate, S=severe. Record location on roof plan.)

Wind Damage _____	Hail Damage _____	Heavy Roof Traffic _____
Vandalism _____	Debris _____	Mechanical Damage _____
Cracks _____	Punctures _____	Deteriorated Felts _____
Fishmouths _____	Blisters _____	Alligatored Coating _____
Splits _____	Buckles _____	Standing Water _____
Exposed Felt _____	Low Spots _____	Other _____

Figure 3-13. *(Continued)*

FLASHINGS:

General Condition: Good _____ Fair _____ Poor _____

Watertightness: No leaks _____ Leaks during continued rain _____

Leaks every rain _____ Leaks during high wind _____

Leaks continuously _____

Check for evidence of: (indicate level of damage as L=light, M=moderate, S=severe. Record location on roof plan.)

Base Flashings:

Deteriorated Base Flashing _____ Open Vertical Joints _____

Flashing Separated from Wall _____ Sagging Base Flashing _____

Deteriorated Surface Coating _____ Missing Counterflashing _____

Punctured Base Flashing _____ Cracked Felts _____

Insufficient Flashing Height _____ Movement _____

Gravel Stop:

Deteriorated Stripping Felts _____ Deteriorated Metal _____

Flashing Separated from Wall _____ Loose Flashing Flange _____

Drains:

Standing Water around Drain _____ Clogged Drain _____

Deteriorated Stripping Felts _____ Deteriorated Metal _____

PHOTOGRAPHIC RECORD:

Prints _____ Slides _____ Video _____ Other _____ None _____

Attached _____ Other Location _____

(NOTE: Each photo/tape should be identified with Building ID, inspection date, inspection form number and a description of what is being shown.)

ROOF PLAN: Draw a roof plan on the next page showing the location of all problem areas found. Also note any changes or additions to the roof since the roof was first completed.

Form filled out by: _____ Date: _____

INSPECTION RECORD: (File this inspection form with the historical record.)

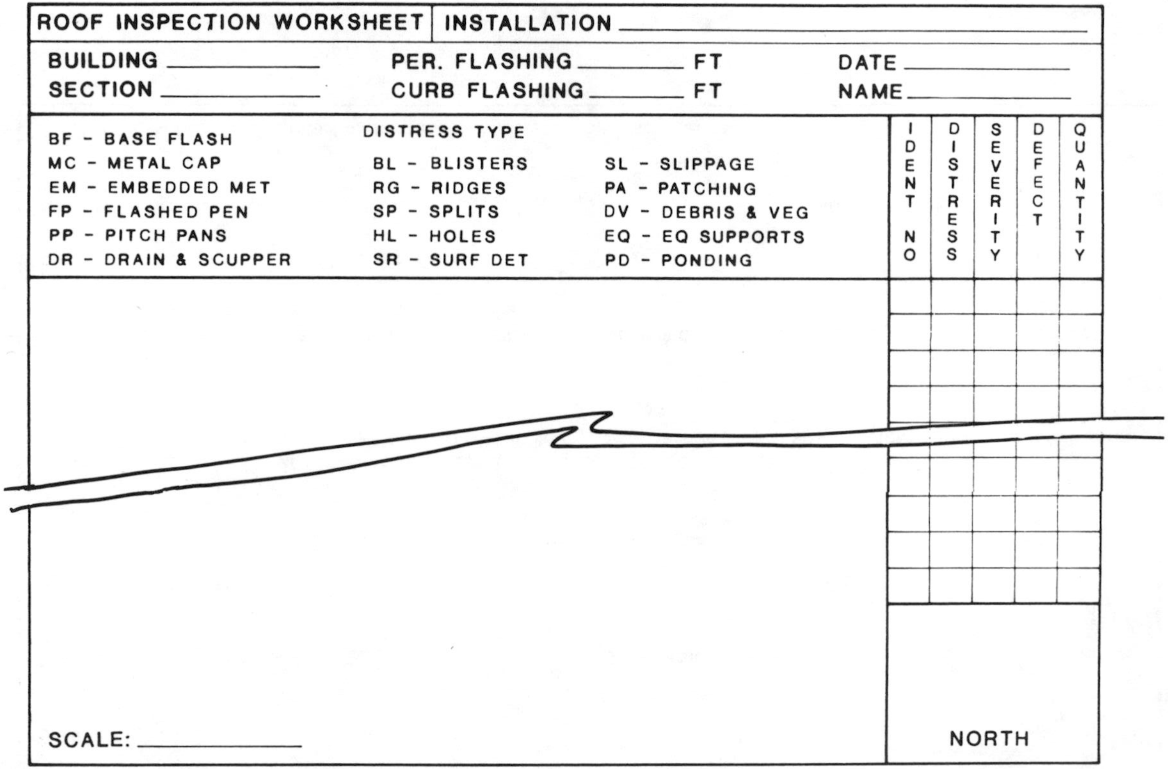

ROOF INSPECTION WORKSHEET | INSTALLATION _____

BUILDING _____   PER. FLASHING _____ FT   DATE _____

SECTION _____   CURB FLASHING _____ FT   NAME _____

DISTRESS TYPE

BF – BASE FLASH   BL – BLISTERS   SL – SLIPPAGE

MC – METAL CAP   RG – RIDGES   PA – PATCHING

EM – EMBEDDED MET   SP – SPLITS   DV – DEBRIS & VEG

FP – FLASHED PEN   HL – HOLES   EQ – EQ SUPPORTS

PP – PITCH PANS   SR – SURF DET   PD – PONDING

DR – DRAIN & SCUPPER

IDENT NO	DISTRESS TYPE	SEVERITY	DEFECT	QUANTITY

NORTH

SCALE: _____

Figure 3-14. Roof inspection worksheet from *Membrane and Flashing Condition Indexes for Built-Up Roofs, Volume II: Inspection and Distress Manual*. (Courtesy: U.S. Army Corps of Engineers, Construction Engineering Research Laboratory)

62

3. When repairing damaged areas, the work should be extended beyond the damaged areas.
4. Reinforcement should always be used when repairing tears or punctures in the roofing membrane.
5. Before applying asphalt coatings the areas should be primed. (Priming is not necessary for pitch applications.)
6. Potential problem areas should be well marked for future observation.

In many cases, time does not allow for a complete job of repairs, and maintenance and emergency measures are necessary. If this is the case, coat damaged areas with roofing cement, inbed felts, or reinforcing fabric and recoat the area. Follow similar emergency procedures for base flashings.

3.5.1 Drainage

Drainage problems are caused by inadequate slope and/or blocked drainage (Figs. 3-15 and 3-16). Generally, drainage cannot be changed after the roof is constructed without extensive and expensive modifications. Installation of additional drains, however, is possible. If roof drainage is a problem, frequently inspect areas where ponding occurs and arrest any roofing deterioration without delay. Many roofing problems result from the lack of simple and inexpensive maintenance procedures. Neglecting to keep the roof clean and drains open makes the roofing system highly susceptible to early failures.

3.5.2 Erosion

Areas void of gravel and bitumen are another frequent occurrence on BURs. Wind is the main cause of erosion of the roof surface. On graveled roofs, the wind can scour away the gravel and the flood coat, exposing the felts and making them susceptible to moisture. On smooth, surfaced roofs, the coating can also be worn off by wind exposing the felts. Eroded areas are more pronounced at corners where the wind action is most severe. If not properly maintained, in time these naturally weathered areas can cause significant problems.

Repair the bare roof surface areas by recoating smooth-surfaced roofs and by recoating and regraveling graveled roofs. If felts are deteriorated, repair them before new coatings are applied. Prime areas that are to receive asphalt coatings. (The hot bitumen, generally, does not adhere to the unprimed areas and allows water to seep under the new coating.) Primer is not required for the areas to be coated with coal tar pitch.

Figure 3-15. Broken drain cover can cause obstruction by allowing debris to enter the drain.

Figure 3-16. Blockage of drains can be due to accumulation of debris on the roof. The illustration also shows alligatoring.

3.5.3   Alligatoring

Alligatoring is another BUR surface condition that requires attention (Fig. 3-17). Its cause is primarily sun exposure. The sun oxidizes the unprotected asphalt coating, which contracts into small areas separated by crevices. If alligatoring is not stopped, these crevices can penetrate to the felts and allow water to enter the roof membrane. Generally, the thicker the coating, the greater the stresses exerted on the membrane by the alligatoring action.

Figure 3-17. Alligatoring can expose felts to water, causing their degradation and eventual failure. Pipe supports dig into the roofing.

To correct alligatoring, clean the surface, prime it if necessary, and apply one of the following coatings: clay asphalt emulsion, bitumen, or if the roof is smooth-surfaced, roof-reflective asphalt-based aluminum coating. If clay asphalt emulsion is used, the roof has to have positive drainage. Otherwise, the clay emulsion coatings will rapidly deteriorate. As already mentioned, increasing the thickness of the coating can frequently increase alligatoring. Industry standards should be referred to for proper coat thickness.

3.5.4 Felt Slippage

Felt slippage (Fig. 3-18) can be attributed to the use of improper bitumen, too much bitumen, excessive slope, or a combination of the three. If slippage is allowed to progress, deformation of felts and lack of adequate coverage over the roof can result.

Corrective measures include fastening of plies if slippage is not severe (it is important to ensure a seal around each fastener). If the slippage is extensive, and felts are deformed, they must be replaced. Of course, factors causing slippage should be corrected to prevent the problem from recurring.

3.5.5 Blisters

Blisters could contribute significantly to a premature failure of the roofing system (Figs. 3-19 and 3-20). Blisters are caused mainly by entrapped and expanding moisture directly under or within the roofing membrane. The

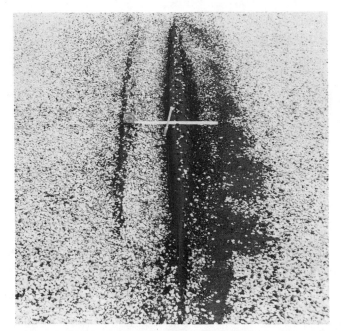

Figure 3-18. Felt slippage reduces roof coverage by felts.

cause of moisture within the plies is the lack of adhesion between layers of felts. Gaseous emulsions from chemical insulations also cause blisters.

Every maintenance inspection of the roofing system should include making a close examination for blisters. Thus, at least one of the roof inspections every year should be conducted on a hot day when blisters are most pronounced. Blisters vary in size from less than 1 foot in diameter to very large ones. Although small blisters do not require any urgent maintenance, the blistered areas are weaker and are susceptible to punctures. Large blisters may cause delamination of felts over substantial areas. They can make the roof surface highly susceptible to physical damage such as rapid wear from wind scouring and interfere with drainage. When blistered areas are repaired, determine the source of the problem or blisters will reappear. Repair procedures include cutting and drying out the blistered area, and repairing it with alternate layers of bitumen and felts.

3.5.6   Splits

Splitting of BUR membrane can spell a disaster (Figs. 3-21 and 3-22). Whereas eroded or blistered roofing areas can be detected before water enters the building, splits are usually discovered after some water damage has already

Figure 3-19. Although roof blisters generally are not a serious problem initially, they are susceptible to wear and puncture.

Figure 3-20. Blister failure can cause instantaneous water damage due to splitting. *(Courtesy: Division of State of Wisconsin Facilities Management.)*

occurred. Splits are caused by the lack of expansion joints, excessive building movement, concentrated loading, or extension of cracking of substrate through the roofing membrane. Splits have also been caused by ice action on the roof.

The key to eliminating splitting is determining the cause and rectifying the problem at its source. You may have to install additional expansion joints or make other major changes. To repair splits, strip in felt plies, and cover with

Figure 3-21. Splits of roofing felts can cause severe damage to roofing and other building systems.

Figure 3-22. Roof splits can cause disaster without warning. Frequent roofing inspections are essential to avert any potential roof defects such as splits.

bitumen and gravel if the roof is graveled, or coat with bituminous coating if the roof is smooth-surfaced.

3.5.7 Ridging

Linear blisterlike deformations of roofing membrane, known as ridging, are caused by moisture condensing on the underside of felts over the joints of insulation boards (Fig. 3-23). Single layers of insulation readily allow moisture to migrate through the joint to the underside of felts where it can condense. Bending up of insulation boards at the edges will also produce ridges, as will the membrane that slips and wrinkles. If more than one layer of insulation with staggered joints is used, this problem is minimized. Roof failures caused by ridging of felts are similar to roof failures caused by blistering: The surface tends to erode and becomes susceptible to physical damage. Minor ridging does not generally cause roof failures. Excessive ridging, on the other hand, poses serious problems and must be repaired. To repair ridges, cut out and strip areas, using felt layers and appropriate bitumen.

3.5.8 Fish Mouths

Fish mouths occur along the edges of felts (Fig. 3-24). These defects can be caused by deformed felt edges or lack of adhesion in these areas. Fish mouths should be repaired without delay for they tend to allow water to enter the felts. To repair this condition, cut fish mouths to lie flat and attach them to the roof with mastic. Place additional felts and mop appropriate bitumen over the area being repaired.

3.5.9 Punctures

Physical damage to the roof includes punctures that may be caused by the abuse of roofing by users of the roof surface. Penetrations through the roofing membrane can also develop from backing out of fasteners, hail, and eroded-through areas over blisters and over ridges. Unless punctures are detected right after they are made, much damage can occur before the defects are discovered. Punctures result in definite roof leaks that can cause wet insulation and substrate and subsequently even damage to the building interiors. When repairing punctures, ensure that the materials within the roofing system, as well as the materials used for repairs, are dry. Wet insulation and damaged felts should be replaced. Repair punctured areas by cleaning and stripping in felts with appropriate bitumen. Take care to include an area much larger than the damaged area itself when repairing punctures.

Figure 3-23. Ridging, like blistering, makes roofing membrane susceptible to accelerated erosion and puncturing.

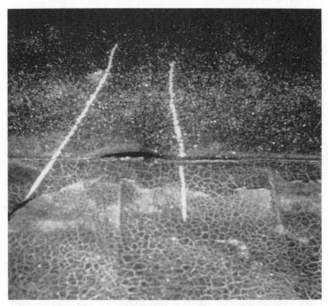

Figure 3-24. Fishmouths expose unprotected felts, allowing water to enter through the opening.

3.5.10 Flashings

Base flashing failures occur in the form of open vertical laps (Fig. 3-25), punctures (Fig. 3-26), flashing separation from the wall (Fig. 3-27), and flashing coating deterioration. Corroded or loose metal counter flashings and wall cappings can also contribute serious roofing failures (Fig. 3-28). Since roofing membranes contract but do not return to their original size, base flashings pull away from the vertical surfaces such as parapets, making these areas highly vulnerable to physical damage. Modified bitumen membrane can be used for base flashings in these areas to provide the necessary strength.

Repair of damaged flashings should generally follow roofing membrane repair procedures. When repairing open laps in base flashing, adhere the flashing with flashing cement, embed reinforcement over the lap into the cement, and recoat the area with flashing cement. If flashing has deteriorated, or if roofing materials are damaged, replace the damaged components before making final repairs. If materials are wet, allow them to dry out. When repairing punctures in the base flashing make sure the damage caused by the water entry is corrected first. To repair base flashings that have separated from the wall, partially remove the counterflashings to allow access to the base flashing. The base flashing should be reattached with flashing cement and mechanical fasteners where appropriate. The counterflashing should then be appropriately reinstalled.

During roofing inspection, flashing conditions should always be carefully examined. Any areas of base flashing that show signs of wear should be cleaned and recoated. Problems relating to metal counterflashings, metal cappings, and root penetrations include improper detailing and installation (Figs. 3-29, 3-30, and 3-31 on pages 75 and 76), physical damage, loss of seal (Fig. 3-32 on page 76), and fastener backout (Fig. 3-33 on page 76). Most of the failures resulting from the deterioration of these elements can be easily prevented if time is taken for regular inspections and maintenance. If these metal components are deformed they should be repaired or replaced. A counterflashing that does not adequately protect the base flashing makes the base flashing susceptible to failure. Recoat corroding flashings and cappings. Replace backing-out fasteners with larger fasteners.

3.5.11 Pitch Pans

Pitch pans are a major maintenance problem (Fig. 3-34 on page 77). If a building has these pipe installations, they will require frequent and regular inspections so that PM work can be performed on them as often as necessary. Widely varying exterior temperatures and other influences tend to dry out,

Figure 3-25. Open lap in base flashing—a potential water problem. *(Courtesy: Division of State of Wisconsin Facilities Management.)*

Figure 3-26. Punctures in base flashings provide ready access to water. Backing, such as cant strips, minimize the occurrence of puncture.

Figure 3-27. Flashing separating from the wall presents a serious problem requiring immediate attention.

crack, shrink, or otherwise take their toll on whatever material is in the pitch pan. Pitch pans are not cost-effective installations, and for better results they should be substituted with proper sleeves and cone units around and over the pipes as illustrated in *The NRCA Roofing and Waterproofing Manual*. Pitch pans should never be used on multiple pipes adjoining each other. Special enclosures should be provided to keep the complex of pipes totally weathertight.

3.5.12   Gravel Stops/Roof Edge Details

Another weak area in the roofing system is the gravel stop or the edge of the roof. For best results, gravel stops are not advocated. Since the coefficient of

Figure 3-28. If corrosion is not arrested, the capping will corrode through, allowing water to enter the wall and roof systems.

expansion of the metal and the roofing membranes is substantially different, the metal gravel stop tends to split the stripping in felts (Fig. 3-35). While splits may not occur at every joint in the metal, it is likely that splitting of felts will take place, especially on long runs of metal edging. If gravel stops are used, strip in edges with two plies of Type IV fiberglass felts and one ply of a modified bitumen membrane. Inspect the roofing edge frequently and repair splits as necessary. Of course, corroded or deteriorated gravel stops or roof edgings should be cleaned and repainted or replaced.

3.6   SUMMARY

BUR systems (felt-mopped together with asphalt or coal tar pitch) have been in existence since the mid 1800s. Properly designed, constructed, and maintained, these roofs can provide many years of satisfactory service. Design, construction, and maintenance designate areas that play a critical role in the performance of BUR systems. Without proper PM, even a well-designed and properly constructed roof will fail relatively early. Likewise, no amount of PM will correct a poorly designed and constructed system.

If the many benefits of PM are to be realized, individuals must know the

Figure 3-29. Improperly designed and installed counterflashing.

Figure 3-30. Unflashed pipe penetration—a potential problem. *(Courtesy: Division of State of Wisconsin Facilities Management.)*

Figure 3-31. Condition susceptible to water entry.

Figure 3-32. Seal failure in wall capping. *(Courtesy: Division of State of Wisconsin Facilities Management.)*

Figure 3-33. Backed-out fasteners result in inadequate attachment of counterflashing, causing water entry through the joints.

Figure 3-34. Failed pitch pan. Pitch pans are poor solutions to roof penetration problems.

function of the roofing system components, materials and characteristics of the components, and methods of constructing the system. Ill-informed attempts to correct a failed area by applying an asphalt coating is not only inadequate for preventing failures or making repairs but also a wasteful use of resources.

Maintaining roofing systems or making repairs involves working with the materials that are in place. Any materials used in maintenance and repair must conform to the existing conditions. BUR failures must be understood. Is the problem caused by poor drainage, erosion of bitumen, felt slippage, splits, flashing failure? Without knowing the cause of degradation, defense against the attack of weather penetration, freezing wind, sun, and so forth cannot be effectively provided.

Since PM, unlike new construction, deals with the systems and components in place, maintenance personnel must have the skills to analyze problems and to prescribe remedial action. In many instances, these decisions need to be made without delay in order to prevent disastrous effects caused by leaks.

Some failures occur without any warning. They can be highly damaging not only to other roofing components but to the interiors of the building as well. Thus, PM must include continuous monitoring of these high-risk areas.

Finally, performing the actual maintenance and repairs requires a high degree of expertise. Again, unlike new construction, in many instances BURs and their components must first be disassembled before correction is possible,

Figure 3-35. Failed roof edge. Result of using materials with different coefficient of expansion.

as in the case of base flashing repair. Special skills are required that many individuals dealing only with new construction do not have. BUR is a time-proven roofing system that can work satisfactorily for many years, provided certain conditions are met. A well-organized and well-executed PM program is one of these requirements.

3.7   SUGGESTED READINGS

Baker, Maxwell, C. 1980. *Roofs: Design, Application and Maintenance.* Montreal, Quebec, Canada: Multiscience Publications Ltd.

Griffin, C. W. 1970. *Manual of Built-Up Roofing Systems.* New York: McGraw-Hill.

The NRCA Roofing and Waterproofing Manual, 3rd ed. 1989. Chicago, IL: National Roofing Contractors Association.

Roof Maintenance. 1989. Englewood, CO: The Roofing Industry Educational Institute.

3.8   ASSOCIATIONS WITH INFORMATION ON BUILT-UP ROOFS

American Society for Testing and Materials (ASTM)
1916 Race Street
Philadelphia, PA 19103

Asphalt Roofing Manufacturers Association (ARMA)
6288 Montrose Road
Rockville, MD 20852

Cold Regions Research and Engineering Laboratory (CRREL)
72 Lyme Road
Hanover, NH 03755

Factory Mutual (FM) System
1151 Boston-Providence Turnpike
Norwood, MA 02063

Institute of Roofing and Waterproofing Consultants
51 West Seegers Road
Arlington Heights, IL 60005

Midwest Roofing Contractors' Association (MRCA)
8725 Rosehill Road, Suite 210
Lenexa, KS 66215-4611

National Roofing Contractors' Association (NRCA)
6250 River Road
Rosemont, IL 60018

The Roofing Industry Educational Institute (RIEI)
14 Inverness Drive East
Building H, Suite 110
Englewood, CO 80112-5608

Steel Deck Institute (SDI)
P.O. Box 3812
St. Louis, MO 63122

Underwriters Laboratories, Inc. (UL)
333 Pfingsten Road
Northbrook, IL 60062

4. Single-Ply Roofing Systems

Rene M. Dupuis

4.1 INTRODUCTION

An attempt to use single-ply membranes for roofs in the United States was made in the mid-1960s. By the late 1970s, single-ply roofing began to make an impact on the industry and had garnered approximately 20% of the entire market. By 1985, the single-ply roofing industry had approximately 50% of the entire low-sloped roofing market in the United States. If a building was constructed in the 1960s, it is more than likely that it has a built-up roofing (BUR) system. If it was constructed in the 1970s or 1980s, it is highly possible that the building has a single-ply membrane for the roof.

The development of single-ply roofing systems primarily began in Europe. The European single-ply systems matured in use before they were imported to the United States. Single-ply roofing exhibited distinct advantages over the traditional BUR system. The installation of single-ply sheets proved to be very quick and clean as compared to the installation of BUR. Also, in many cases, the single-ply sheets could be installed over the existing BUR membrane without removing it, which realized substantial cost savings.

In the 1960s, U.S. designers, builders, and manufacturers did not understand single-ply technology, especially as it related to the installation of the systems. The manufacturers did not completely understand the performance standards that a single-ply sheet roof membrane had to meet. The application technology was not clearly thought out. Numerous sheet products were improperly formulated or compounded.

Today, the problems of the 1960s have been largely resolved. The skills and technology are available for the production of dependable watertight roofing using single-ply membranes. The roof, however, must not only be properly constructed, it must also be maintained.

Achieving a high-performance single-ply roof is a demanding undertaking. Presently more than 80 manufacturers provide single-ply products. Since it is safe to estimate that each of these manufacturers has 5 to 8 different specifications, there are some 500 roof specifications dealing with single plies that are available for use on the low-sloped roof. This variety tends to confuse building owners as well as roofing experts. Thus, it is vital that a building owner document exactly what single-ply system has been selected for the building and understand the technicalities of membrane properties, anticipated weathering behavior, and maintenance needs.

Workmanship on single-ply systems is another key issue that has to be dealt with effectively to obtain a problem-free roofing system. In comparing the workmanship needs for a single-ply roof with those of a conventional BUR, there are distinct differences. The most glaring difference is that the applicator of a single-ply roofing membrane must be especially meticulous and precise, since single-ply membrane and one lap joint are the only difference between keeping the building dry or suffering water damage. In single-ply systems there is no safety factor of multiple weatherproofing plies as there is in traditional BUR systems.

The preventive maintenance (PM) of single-ply roofs has become very important. Because of the lack of PM on many single-ply roofs, numerous single-ply roofing failures have occurred prematurely. Preventive maintenance cannot be effectively accomplished on any building system unless the physical and chemical properties of building materials are understood. Skills must be developed to properly install these materials. Periodic and regular inspections are essential. The inspector must know where to look for potential problems, and once these problem areas are identified, the proper steps must be taken for repairing and maintaining these problem areas. Because of the wide variety of single-ply membrane materials and components, this chapter will discuss the properties of these materials before embarking upon PM and repair considerations.

4.2   TYPES AND COMPOSITION OF SINGLE-PLY MEMBRANES

Anyone involved in PM of single-ply roofing should know the performance capabilities of single-ply membranes and the advantages and disadvantages of using different products in different situations.

Not all materials can resist the environmental influences equally well. Single-ply membranes differ substantially in their ability to withstand weathering, ponding, heat, oil, petroleum products, and ultraviolet rays. Single-ply membrane installations, especially the earlier ones, have met with varying degrees of success. In some cases, wrong materials were used; in other instances, membranes did not perform as expected. Today many of these installations require intense maintenance and repair. Often PM personnel need to make decisions whether to maintain/repair the existing system or replace it with fewer materials that have a better performance record. Another area of concern is that the importance of materials' compatibility should never be underestimated. It is most critical in PM work where one has to deal with existing construction. Using incompatible materials generally spells an almost immediate disaster.

Finally, decisions on PM often need to be made without any delay. Whereas in new construction there generally is time for product investigation, many decisions dealing with PM have to be made immediately. Under these circumstances the knowledge of materials becomes invaluable.

4.2.1 Chemical Background

A number of basic chemical concepts will help in the understanding of distinctive behaviors and differences of each single-ply material. All plastics and synthetics (or natural rubbers) are members of a large organic chemical family known as polymers. With plastic and rubber materials, the molecules contain primarily carbon, hydrogen, chlorine, or other atoms in very specific arrangements. This special group of molecules is referred to as monomers.

Polymers are made from the chemical union of five or more identical units of monomer. Chemists can link several different monomers to produce many different types of polymers. Many synthetic polymers are well-known; ethylene, for example, is used as a monomer to produce the familiar plastic material polyethylene. In this instance, polyethylene is a polymer. Vinyl chloride (the monomer) can be made into a polymer, polyvinyl chloride (PVC). Varieties of synthetic polymers are therefore possible, allowing use of these materials in a wide range of application. Consider the material's strengths and weaknesses if it is to be used as a roof membrane.

Although these polymers are the basic materials used to produce roof membrane sheets, additional (and very important) materials must be added to the compound before the sheet can be successfully made. Normally, however, the sheet is identified according to the majority polymer that is present, such as PVC for polyvinyl chloride and EPDM for ethylene propylene diene monomer.

4.2.2 Thermoplastic Versus Thermoset Process

Single-ply roof sheet materials are generally referred to as plastic or elastomeric. A plastic material is capable of being shaped or molded by the application of heat and/or pressure. Materials of this nature are normally referred to as thermoplastics. Materials such as PVC are thermoplastic and can be heat- or solvent-welded. Polymer chains are linear and, with the aid of plasticizers, they are flexible. Heat allows them to slip over each other and bond or weld as required. Thermosets or elastomerics are polymer materials that have been cross-linked with their polymer chains intertwined and not able to slip from each other. Thus, if you pull on a thermoset material (such as a piece of rubber) it snaps back. Thermosets are referred to as cross-linked, cured, or vulcanized materials and cannot be reshaped, molded, or welded. Another material in use is an uncured elastomer. This material can be cured or vulcanized.

Many of the single-ply systems are composite membranes, having reinforcing fabric and several different materials within a sheet. The materials most commonly used for single ply roofing membranes in the United States today are described below.

4.2.3   Chlorinated Polyethylene (CPE)

This thermoplastic sheet is composed of high-molecular-weight, low-density polyethylene that has been chlorinated to a given level. Plasticizers may also be added to PVC. This sheet material offers excellent weathering resistance and chemical resistance to acids. The sheet is normally available in gray or white.

4.2.4   Chlorosulfonated Polyethylene (CSPE) (Hypalon)

Hypalon is basically a chlorinated polyethylene containing chlorosulfonyl groups with a high-molecular-weight, low-density polyethylene. This material is cured and has an outstanding resistance to ozone, heat, and weathering. It has fairly good resistance to general chemical attack, but will swell in chlorinated and aromatic-type solvents. It can be formulated in a wide variety of colors; color stability is reported as good. This sheet material may contain a backing; it has been used in Europe and the United States.

4.2.5   Ethylene Propylene Diene Monomer (EPDM)

EPDM is synthesized from ethylene, propylene, and small amounts of diene monomer. It is similar to butyl rubber but has better weathering resistance. It also has good resistance to corrosive chemicals, diluted mineral acids, and vegetable oils, but is not resistant to petroleum products. This cured elastomeric sheet is available in both plain and reinforced styles; normally sheets are black.

4.2.6   Modified Bitumen

Modified bitumen describes a broad class of materials (and blends) that can be used to change the performance characteristic of a roofing-grade asphalt. In some instances, rubberizers are added; polypropylene, tack oils, or other additives may also be used. Since each modified bitumen composite sheet system is somewhat different, we cannot uniquely describe each one under the generic scheme used here. It is felt, however, that as the use of modified bitumen composite sheet continues, there will be a distinction between those systems using a rubberized approach versus polypropylene modified. Modified bitumen may contain a slate granule topping, aluminum skin, polyethylene surface sheet, or it may be left exposed, depending on the manufacturer.

4.2.7   Polychloroprene (Neoprene)

Neoprene is the generic name for polymers of chloroprene; this material has been used for elastomeric roofing for some time. This cured synthetic rubber sheet has good resistance to petroleum oils, solvents, and weathering. Some formulations require a protective coating of chlorosulfonated polyethylene (Hypalon) to retard weathering. The weathering grade of neoprene is black; nonweathering grades may be lightly colored and should be protected from sunlight.

4.2.8   Polyisobutylene (PIB)

PIB is a synthetic rubber sheet classified as an uncured elastomer composed of isoprene, highly molecular isobutylene, carbon black, and aging protectors. This material has good ozone, ultraviolet, and weathering resistance. The sheet is normally available with a nonwoven polyester fleece backing.

4.2.9   Polyvinyl Chloride (PVC)

PVC is a polymer synthesized from vinyl chloride and is thermoplastic in nature (can be fused or reshaped by the application of heat or a solvent). Through proper formulation with plasticizers and stabilizers, PVC roof sheet can exhibit elasticlike properties and resist weathering. This material is available in both a reinforced and plain version. It is resistant to acids, alkalis, and many chemicals, and is available in gray or tan.

4.3   PHYSICAL PROPERTIES OF SINGLE-PLY MEMBRANES

Tensile strength, ultimate elongation, hardness, and mil thickness are some of the physical properties measured on most of the single-ply materials. Generally speaking, single-ply membranes are capable of high elongations (greater than 200%) at room temperature. With the exception of modified bitumen systems, single-ply materials are normally made available in sheet thicknesses ranging from 32 to 60 mils.

There has been some debate about the effectiveness of the reinforcing fabric within the single-ply sheet. If a sheet does contain reinforcing, its ultimate elongation will not meet that of a homogeneous (plain) sheet. The question should not necessarily be which one has the highest elongation but which sheet material can absorb more strain before losing its watertight integrity. When a reinforced single-ply sheet ruptures its internal fibers, however, it may not necessarily lose its watertight integrity.

The modified bitumen composite sheet systems are generally thicker than other type sheets, ranging from a nominal thickness of 60–160 mils. The tensile and elongation properties of modified bitumen composite sheets are in some instances parallel to the elastomeric or plastic sheets, although modified bitumen has a wider range of ultimate tensile strength. This variation is primarily due to the type and amount of reinforcing contained in the modified bitumen composite. Although it is still not known how much elongation and tensile strength are actually required, this wide-range capability is important to the performance needs of the roof membranes.

The ability of the single-ply membrane to breathe or transmit water vapor is debatable. Current research findings do not substantiate that the single-ply membrane breathes. Although many laboratory tests have been done on the rate of water vapor transmission, these tests were, for the most part, run at room temperature. Since a single-ply membrane exposed to the elements will undergo a drastic temperature change, it follows that the vapor transmission rate of single-ply membranes could be affected by the temperature of the membrane. In addition, surface materials such as rock, dirt, or ice could preclude the membrane from effectively transmitting vapor.

Reroofing projects covering moisture-laden substrates are sometimes cited as having dried out after a period of time. The argument is that the single-ply membrane allows the moisture to escape through the membrane. Close inspection of some of these field installations shows that downward drying may inadvertently take place. Loose-laid, single-ply membranes also allow for the lateral transmission of water vapor that in turn escapes along adjacent parapet and perimeter walls. In summary, there is little current evidence that a single-ply membrane laid over a wet or moisture-laden substrate would then allow it to dry by vapor transmission.

Dimensional stability is often cited as another area of concern for single-ply roofing. In general, any reinforced single-ply sheet exhibits a higher degree of dimensional stability than an unreinforced sheet. These sheets generally exhibit a 0.5–1% range of shrinkage in the dimensional stability test, depending on the type of reinforcement. Unreinforced or single-ply sheets can in some instances undergo a shrinkage in excess of 3 or 4% in a dimensional stability test. Thermoplastic or PVC sheets would, by their nature, exhibit a higher degree of dimensional change (when exposed to high temperature) than a substantially cured thermoset (synthetic rubber sheet membrane). Currently, manufacturers are aware of this potential problem and are monitoring production.

Heat aging or prolonged exposure of single-ply products to high heat is of concern to many single-ply manufacturers. Rooftop temperatures below 110° F do not pose a great danger for most single-ply materials. However, the effects of prolonged exposure at temperatures in excess of 150° F should be evaluated by the manufacturer.

It does not necessarily follow that a material successfully used in a mild or northern climate will perform in a hot or humid environment such as the southern or southwestern areas of the United States. Research by manufacturers and others continues to evaluate the performance of the single-ply sheets under various conditions of high heat exposure over time.

The primary concern for low-temperature behavior with single-ply membrane systems is the ability for the material to flex. Some single-ply materials cannot be installed below 40° F whereas others may be installed at much lower temperatures. A specifier of any single-ply system should be concerned with cold weather installation. While the materials may withstand the rigors of cold temperature application, human discomfort may not allow for the quality of workmanship required for assembling the system.

A number of other physical and chemical properties of single-ply systems are being investigated: water absorption, coefficient of expansion, tear resistance, wind uplift. The challenge always remains to ensure that the test conducted is appropriate for answering the technical question at hand: How does this lab test compare to the membrane performance on the roof?

4.4 TYPES OF SINGLE-PLY ROOFING SYSTEMS

Single-ply roof membranes are normally installed in one of the following assemblies: loose laid and ballasted, partially attached or mechanically fastened, totally adhered, or protected membrane roof. Each of these roof assemblies offers distinct advantages and disadvantages; some are readily apparent whereas others are not as obvious.

4.4.1   Loose-Laid and Ballasted

These systems are generally easy to install, but may require special considerations at points of flashing attachment. The membrane has to be ballasted to prevent wind blow-off; a nominal 10 lb-ft^2 of stone ballast (or more) is usually required (Fig. 4-1). This extra dead load on the roof structure should be taken into account by the building designer.

Quite a few structures do not have sufficient load capacity to allow additional dead loads on the structure and still provide ample live load capacity to carry snow loads according to the building code. Due to the additional load of ballast, deflection calculations should be made to assure that ponding will not become a problem. Loose-laid membranes enjoy the freedom of independent movement and total isolation from any structural settlement or movement. Depending on the substrate type, a slip sheet or vapor barrier may have to be installed over the insulation board or old roof membrane. The stone ballast should be a well-rounded gravel and not contain any sharp or jagged edges.

4.4.2   Partially Attached or Mechanically Fastened

This type of single-ply roof system depends on strip or point attachment methods. Currently, there are a number of techniques for achieving partial attachments of single-ply membranes by the use of mechanical fasteners (Figs. 4-2 through 4-5). Linearly attached systems utilize light-gage steel bars hidden in each lap. Point attachment is achieved by stress plates and metal fasteners or wide plates (12 × 12 in.) that are fastened to the deck, with the single-ply membrane being adhered or bonded to these individual plates. Be cautious when selecting fasteners for the substrate over which they will be used. Trapped moisture, in very small amounts, would not normally be a

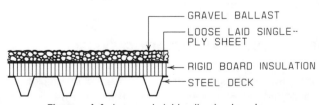

Figure 4-1. Loose-laid ballasted system.

problem to such a system. Lateral venting from underneath a single-ply membrane to exterior wall flashings may be achievable; the same would hold true for a loose-laid system. However, larger amounts of trapped moisture can be a problem since corrosion of fasteners leads to premature failure of the roof system.

The mechanically fastened systems have to ensure that movement or displacement in the vertical direction, due to insulation board compaction, settlement, and other related settlement behavior, do not cause fasteners to loosen or cause them to protrude against the membrane. Since most of the partially attached systems use a metal strip or disc, there is potential for flutter fatigue. Thin elastic membranes could, under high wind conditions and cold temperatures, experience a ballooning or lifting away from discrete attachment points and could rupture after many cycles.

4.4.3 Totally Adhered Systems

These single-ply systems depend on contact cement, cold adhesives, or hot asphalt to uniformly adhere the membrane. Totally adhered single-ply membranes (except those using hot asphalt for adhesion) have been known to lose small areas of adhesion. By and large, these are viewed as cosmetic deficiencies. In some cases, after a warm summer, these membranes readhere themselves. However, if a fully adhered single-ply system is subjected to high winds, and critical areas such as the roof perimeter and corners have lost adhesion, substantial damage can occur. "Peeling" will occur due to the flexibility of the single-ply membrane, allowing the wind to eventually lift and peel away the remaining adhered portions.

4.4.4 Protected Membrane Roof Systems

This system is essentially another form of a ballasted system; in this case, a single-ply membrane is installed loose-laid or adhered to the structure of the roof deck, with the insulation installed over the top of the membrane, protecting it from direct exposure. Ballast is normally laid down over the insulation; paver blocks or other methods of dead loading the insulation board are sometimes used. Tongue and groove insulation boards and tie-down straps also ensure that wind does not get under the insulation and cause damage.

4.5 CAUSES OF FAILURE

Single-ply roof systems are subject to ultraviolet rays, extreme temperatures, chemical contamination moisture, wind, and mechanical damage. A bal-

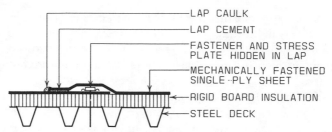

Figure 4-2. Fastener and stress plate hidden in lap. Point attachment is made by each fastener.

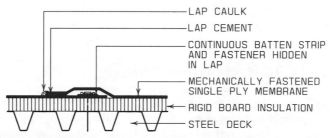

Figure 4-3. Batten strip and fastener hidden in lap. Linear attachment is made by batten strips.

lasted roof protects the single-ply membrane from direct exposure to ultraviolet rays and the extreme ranges of temperature. The ballast provides mechanical protection and, due to its thermal mass, cannot be cooled or heated rapidly. However, ballasted roofs are known to undergo a heavy siltation, depending on how much dirt is in the air. So, although single-ply ballasted roofs do have the advantage of protection from ultraviolet rays and temperature, they may suffer an additional risk due to the presence of soil, bacteria, and water in the spaces between the gravel ballast. Also, locating a leak under a ballasted roof can be time-consuming and difficult.

Fully adhered and mechanically fastened single-ply roofs must withstand direct ultraviolet radiation and temperature extremes. Black-surfaced EPDM roofs have been know to approach 180° F during direct exposure to the sun. Maintenance personnel normally prefer exposed roof membranes because damaged or distressed areas can be more easily found and repaired. Over time, chemical contamination may occur near stacks and rooftop vents that exude hydrocarbons, light oils, and vegetable or animal fats. (Each single-ply membrane manufacturer maintains a list of chemical compounds that are known to cause damage to their membrane.) If a roof vent must be cut in for exhausting air that carries membrane-damaging agents, consider a water trap or gravel box with drainage holes. The trap or box would absorb most of

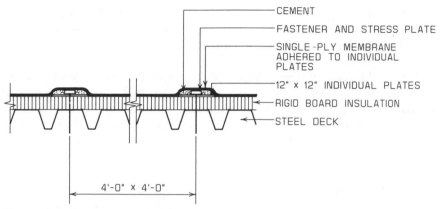

Figure 4-4. Plate bond attachment made by adhering membrane to plates that are 4 ft on center across roof.

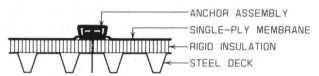

Figure 4-5. Nonpenetrating attachment method using fastened boss under membrane with crimp-fit plastic cap forced over boss, clamping the membrane.

the oils and allow for surface evaporation of the contaminants. Consult the membrane manufacturer for the latest methods of handling different contaminants.

As a rule of thumb, asphalt and coal tar products should not be used on PVC or EPDM membranes. Modified bitumen membranes behave like BUR systems and are also subject to damage by petroleum-based oil products.

Single-ply roof systems are not designed to be walkways for maintenance traffic. Mechanical damage will occur if a single-ply roof is used indiscriminantly for work activities. Protection pads or walkways should be laid down in areas that have predictable needs for traffic or work activity. Some building owners require maintenance personnel to use soft, flotation tires on rooftop carts. They protect the roof from mechanical damage.

High traffic over ballasted roofs can also cause membrane damage since sharp-edged stone will eventually cut the membrane if repeatedly walked on. Rigid insulation board can prevent mechanical damage to a single-ply membrane by absorbing the impact energy from a falling object.

One of the most difficult problems in rooftop maintenance is predicting where the foot traffic will occur, especially in large facilities. Some owners require a rooftop escort on the first trip to the roof to educate the worker on the strengths and weaknesses of the roof system. Virtually every roof manufacturer's warranty excludes damage caused by foot traffic and maintenance personnel.

Due to the wide variety of single-ply membrane materials and different roof systems available, it is imperative to know each roofing component so that maintenance needs are better understood and easier to assess. Also, it is important to understand how these systems fail and what causes their failure.

4.5.1   Substrates

In many instances, roof substrates can contribute significantly to failure. Steel decks are susceptible to rusting if exposed to moisture. This moisture could come from the outside environment or from excessive humidity inside of the building. Under highly concentrated loading, or when low-yield strength of steel is used, the steel deck becomes susceptible to buckling or dimpling. Poured gypsum decks may experience a loss of structural integrity if exposed to the presence of moisture for a lengthy period.

4.5.2   Insulation

The main problems experienced with insulation include shrinkage, expansion, warpage, crushing, and delamination. Moisture and temperature fluctuations are the primary influences causing the shrinkage, expansion, and warpage. Delamination results largely because of the use of incompatible materials in the manufacturing of insulation. Also, certain solvents and adhesives can cause delamination of the insulation sheets. Warpage is due to uneven expansion of the cell structure of insulation boards. A significant problem observed in many single-ply roofing systems is that of crushed insulation. The problem becomes highly critical when crushing of insulation occurs near roof membrane fasteners. Then the fastener protrudes above the insulation, making the roofing membrane susceptible to puncture.

4.5.3   Vapor Retarders

This component fails primarily because of puncturing, tearing, or ineffective lapping of vapor retarder sheets. The actual retarder's effectiveness also depends on thickness and type of product used. Plastic sheets serve as excellent vapor retarders. Bituminous sheets and coated papers are less effective.

4.5.4 Adhesives

Failure of adhesive may occur because of inadequate mixing, poor surface preparation, insufficient amount of adhesive, and leaving the adhesive exposed to the environment for too long before lapping.

4.5.5 Fasteners

There are three main categories of fastener problems: corrosion, loss of holding power, and back-out. Moisture entering the roofing system, whether from outside or inside, will tend to cause corrosion in metal decks and fasteners. If the moisture is not allowed to escape from the old roof when installing the new roof, fastener corrosion will accelerate. When a new roof is placed over the existing roof, fasteners become susceptible to corrosion especially at the interface of the two roofs. The corroding metal deck will substantially or completely reduce the holding power of fasteners as well. On concrete decks, the holding power of fasteners may be impaired if a fastener is set to an inadequate depth into the concrete slab. Moisture is not the only cause of fastener corrosion. In wood decks, for example, fire-retarding chemicals may be the prime cause of corrosion of fasteners.

The loss of holding power by fasteners can also be the result of over-torquing fasteners in steel decks. Wind action and vibration caused by mechanical equipment could cause roofing fasteners to back out. Fasteners that back out have less holding power (providing less anchoring to the roof insulation or single-ply membranes) and become susceptible to membrane puncture (Fig. 4-6).

4.5.6 Flashings

One of the key problem areas on low-sloped roofs is flashing. Due to mechanical forces, wind, and inadequate anchoring, flashing can pull away from the wall, making an area highly susceptible to water entry into the building (Figs. 4-7 and 4-8). Because of mechanical forces, flashing membranes can experience tears or pull out from cap flashings.

4.5.7 Surfacings

Used on single-ply roofing systems, ballast materials could crumble and split due to freeze and thaw cycles. Wind forces could cause the ballast to scour and blow off the edge where a low gravel stop detail is used.

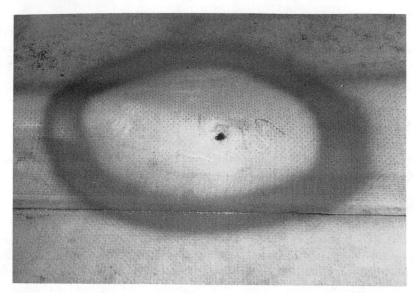

Figure 4-6. Backing-out fastener causes a puncture in single-ply membrane.

4.5.8   Drains

Drains provide controlled removal of water from the roof. A well-functioning drainage system is one of the key factors that ensures an efficiently function-ing roofing system. Many factors cause significant problems to the roof drainage system. Roof drains can be plugged by debris. Mechanical impact can cause cracked drain bowls. Loose clamping of drains can cause substan-tial leaking into the roofing assembly. An inadequate amount of mastic around the drain before it is anchored into place can also be responsible for leaks.

4.5.9   Expansion Joints

Expansion joints are breaks in a building that divide it into separate sections that are free to move individually without affecting other sections. Expansion joints extending through the roof should be covered with a durable, pliable material that allows building sections to move while protecting the building from water entry. Many expansion joint problems are due to improper design and detailing of the joint (Fig. 4-9). Problems occur with the attachment and

Figure 4-7. Counterflashing failure.

Figure 4-8. Improperly attached flashing exhibiting severe failure.

Figure 4-9. Expansion joint failure.

anchoring of the joint cover and also with other aspects such as incomplete joints and jogged joints. Some expansion joint failures can be attributed to inappropriate material selection. Care must be taken to use a material for expansion joints that is compatible with the roofing membranes.

4.5.10   Membranes

In order to design and specify effective single-ply roofing systems, an understanding is necessary of how single-ply membranes fail and what causes failures. Single-ply membranes can be subdivided into four categories, based on their behavior and failure characteristics. The first category includes CPE, CSPE (Hypalon), and PVC. The second category includes EPDM and neoprene. The third category consists of modified bitumen and the fourth category consists of PIB.

Category one, made up of CPE, CSPE (Hypalon), and PVC, can fail in various ways. Membranes may break down due to chemical reactions with materials such as coal tar pitch. Loss or change in plasticizer effectiveness in PVC products can also occur from improper formulation at the time of manufacture. Embrittlement and hardening of the membranes is caused by changes of material properties due to weathering and/or microbial attack.

Delamination of reinforcement is another type of failure of single-ply membranes in this category. It is caused by inadequate strike-through of reinforcement into the membranes. Insufficient cover over the reinforcement causes exposure of reinforcement, which in turn may cause the failure of the membrane. Lap separation, potentially a critical problem, could be caused by inadequate heat or pressure applied during welding, or by contaminated welding solvents. Ponded water, ultraviolet exposure, and/or microbial attack can cause surface crazing of the sheets. Weathering and ponded water can cause loss of lap sealant.

Membranes falling into the second category, EPDM and neoprene, are susceptible to failure through swelling, loss of lap sealant, and separation of laps. Contamination with the various oils or products containing hydrocarbons is generally responsible for swelling problems. Inadequate mixing of adhesive, low-quality adhesive, insufficient amount of adhesive placed in the lap, improper rolling out of finished lap, and improper cleaning of sheets before applying adhesive are causes of lap separation. Loss of lap sealant, of course, like in the previous category, is generally affected by weathering and ponded water.

As indicated above, category three consists of modified bitumen. Surface crazing, enbrittlement, and hardening are problems experienced with this particular product. The cause of these problems, again, is weathering and the presence of ponded water. Improper mix of components (incompatible mix) could also contribute to these types of failures. Delamination could be a problem with modified bitumens if the product has been improperly manufactured. Separation of torched or hot-mopped laps of modified bitumen can be attributed primarily to poor installation practices including insufficient heat or extreme heat used in torch application of modified bitumen or mopping of asphalt that is too cold to achieve adhesion. Finally, as with most other membranes, breakdown of modified bitumen can be seen because of the combination of ponded water and ultraviolet influences.

Category four, PIB, is susceptible to problems primarily in two areas: surface crazing and lap separation. Surface crazing may be caused by ponded water and microbial attack. Failure to properly clean the lap joints, inadequate rolling pressure, and insufficient amounts of solvent used (in the field) to soften PIB could cause serious lap separation problems.

4.6 GENERAL DESIGN CONSIDERATIONS

Poor design, whether in new construction or in maintenance and repair work can lead to high costs and reduced service life of a single-ply roof. Knowledge of the design-caused failures of the roofing systems is essential to the full understanding of why the system is not giving the service expected, or why the system has more repairs and maintenance costs than anticipated. Three basic design factors that can dramatically affect the in-service performance of a roof are (1) slope and drainage, (2) flashings, and (3) design against wind.

4.6.1   Slope and Drainage

Positive drainage has proven to be beneficial to countless roof systems. During the advent of single-ply, many designers and installers felt that positive slope and good drainage did not need to be provided since the new single-ply technology would overcome this issue. Now roof experts have come to realize that any roof system benefits from positive slope. If the roof deck has positive slope and ponding develops, the cause may be that the drain lines are installed next to columns. Consequently, the ponds form between column lines, making roof drainage difficult. A tapered insulation system to overcome localized areas of ponding could be installed if this expense is justified. Or, crickets could be used to direct the flow of water toward the drains. Overflow drains should always be provided since a plugged drain can cause a structural overload. Scuppers should be installed if no overflow drains are present and the roof is surrounded by a raised edge. The scuppers will allow for emergency overflow. Also, consider adding more drains. Before installing the drains, check the existing structural framing to ensure that drain leaders can be installed without major difficulty. Consider the use of drain inserts if the existing drain and associated clamping hardware are cracked or damaged at the roof level.

4.6.2   Flashings

The flashings on single-ply roof membranes are much lighter in construction than conventional BUR flashings. Unfortunately, these details vary so much between single-ply manufacturers that it is difficult to depict a standard detail. Figures 3-3, 3-5, 3-6, 4-10, 4-11, and 4-12 show the difference between BUR and single-ply membrane system's termination at a wall, control joint, and roof edge details. A base and counterflashing detail is generally called for with BURs. Single-ply systems ordinarily call for a simple termination bar to be affixed to the wall at a minimum of 6 or 8 inches above the roof surface. A termination bar must have a heavy bead of caulk to shed water from the wall and over the bar. Unfortunately, caulk shrinks with age and cracks develop with the forces of thermal expansion occurring between the wall and the termination bar. Water will then seep in behind the single-ply sheet and down under the roof.

Termination bars must be checked regularly for conditions of caulk and attachment to the parapet walls. If termination detail appears satisfactory, and the building is still experiencing water problems, check the weep holes in the masonry wall to see that they have not been covered up with roofing sheets.

One design concern of reroofing buildings with single-ply is the ease with which a contractor can install a membrane above the weep holes or a line of through-wall flashing. No amount of maintenance can correct this problem since the weep holes or through-wall flashings will intercept all water coming down the interstices of the wall and direct it under the single-ply membrane.

4.6.3 Design Against Wind

Due to their light weight and variety of attachment methods, some of which are not very effective, wind forces can cause damage or blow-offs to single-ply roofs. Loose-laid roofs with gravel ballast must be checked at all corners for gravel scour. Consider installing a high gravel guard or screen if ballast is being pushed over the edge by wind forces. Reworking of roof edges with larger stone or interlocking pavers may be required.

Mechanically fastened single-ply systems depend on screw fasteners to resist wind forces. They hold the system down to the deck by compression, applied through a stress plate to the underlying insulation board. The corners and perimeter areas should be checked for signs of fastener back-out or fatigue of the membrane itself. High-frequency, low-force winds are unrelenting in their action against a lightweight, mechanically fastened, single-ply roof system. Figure 4-13 shows how a fastener head pushes against the membrane (near a lap splice) when backing out.

4.7 IMPORTANCE OF LAP JOINT CONSTRUCTION

One of the most critical steps in the construction of a single-ply roof is installing the lap joints. This vital step turns individual single-ply sheets into a continuous roof membrane. Before the roof is complete, thousands of lineal feet of successful lap joints will have to be made; a good lap joint must allow for a tolerance dimension during construction.

The roofing contractor must understand the materials that comprise a successful lap joint, how these materials interact, and what temperature conditions are best for making a lap joint. Due to the wide variety of single-ply materials and roof systems now available, the contractor faces a continuous challenge for educating installation crews. Building owners must also know the generic differences.

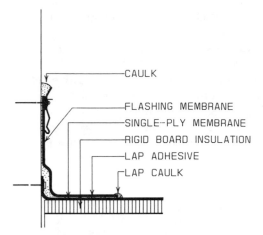

Figure 4-10. Single-ply roof system termination at the wall.

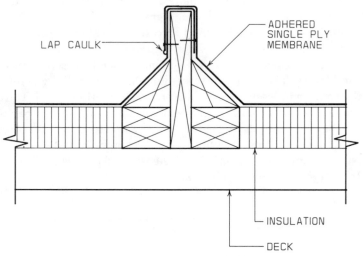

Figure 4-11. Single-ply control joint.

4.7.1   Types of Lap Joints

Because of the wide variety of single-ply sheet materials available, there are many different methods currently used in making lap joints. Generally speaking, lap joints can be:

1. Solvent-welded
2. Heat-welded
3. Made with contact adhesive

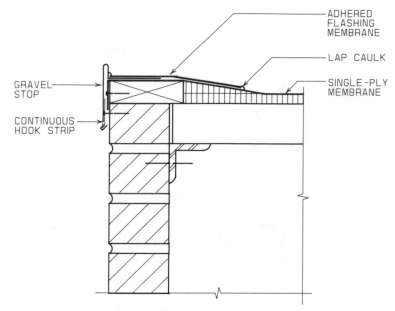

Figure 4-12. Single-ply roof edge detail.

4. Made with hot asphalt
5. Torched (cohesive)
6. Self-adhering
7. Made with tape splices

Each of the methods offers distinct advantages in certain roofing situations.

4.7.1.1 Solvent Welding Method

Seams of thermoplastic membranes, such as CPE and PVC, can be solvent-welded with tetrahydrafuran. Normally, the membrane manufacturer supplies the solvent. It may contain other additives or be a clear fluid.

Often a "bodied" solvent is used, with the parent membrane material partially dissolved in the solvent, having a colored appearance similar to the membrane. Solvent welding can be accomplished with a brush or similar applicator. The technique is quite simple, but the conditions under which a solvent weld can be accomplished are somewhat stringent. There should not be any dirt or moisture on the surfaces of the sheets being spliced together. Any free water, condensation, or dew will defeat the lap splice.

After applying the solvent, the sheets are simply lapped together and pressure is momentarily applied. The lap splice gains strength and becomes mechanically stable normally within minutes. A portable device for solvent

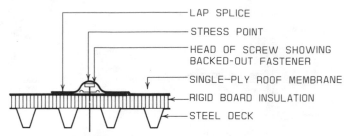

Figure 4-13. Backed-out fastener.

welding allows the mechanic to simply roll the solvent applicator along the side lap, uniformly applying the solvent.

The solvent process, however, may be risky in cold weather, since condensation may occur in minute amounts within the lap as the solvent evaporates. This drawback limits this method during winter construction.

4.7.1.2   Heat Welding Method

Lap seams can be heat-welded at ambient temperatures below 40° F without affecting the lap splice. Both hand-held and motorized heat welders are available and have been in use for some time. Heat welding is accomplished by bringing the surface temperature of the membrane up to its softening point; with the application of pressure, the two materials fuse together, forming a solid weld. A hand-held roller is normally used in conjunction with the heat welder.

Welding by this process takes some experience. Personnel who are new to this technique should be given test strips and flashing samples on which to work. Motor-driven heat welders can give excellent results with high productivity. The machines are set in place and allowed to traverse the roof under their own power. A mechanic guides the device and generally can expect 12–15 ft-min of production. Many manufacturers prefer heat-welded lap joints. In some cases, it is the only approved method.

If motor-driven heat welders are used, the roofing contractor should use portable generators, since long power leads can result in a voltage drop at the welder. Inexperienced crews will unwittingly build a "cold" lap joint if they are not sure of their voltage supply or heat output at the welder.

4.7.1.3   Contact Adhesives Method

Rubber membranes, such as EPDM, use large quantities of contact adhesive to accomplish lap joints. The rubber-based contact adhesive in use today has a solvent carrier. Its application requires coating of both surfaces of the sheet, allowing the solvent to flash off.

Figure 4-14. Single-ply membrane lap failure.

Figure 4-15. Improperly constructed single-ply membrane joint.

The contact adhesive (sometimes referred to as splicing cement) is normally easy to apply, has reasonable storage life, and dries fast. In most cases, however, this type of lap joint is the weakest splice in single-ply roofing technology (Figs. 4-14 and 4-15). The adhesive lap splice cannot offer the strength of a solvent or heat-welded lap, which is due in part to the stubbornness of EPDM as a bonding site.

In addition, the EPDM sheet is covered with a talc dust as part of the vulcanizing process (to prohibit the sheet from blocking or curing on itself in a roll). A successful lap joint then requires two critical steps:

1. All talc must be cleaned from the surface.
2. A uniform coating of contact adhesive should be applied with no air pockets or blisters entrapped within.

Cleaning of the talc normally requires a solvent and clean cloth. Care should be taken in removing the talc. Workmen have a tendency to simply push the solvent-laden cloth along the lap joint, pushing the talc from one end to the other and not actually removing it. Good lap joint preparation for the adhesive splice cannot be overemphasized. Crews should be diligently instructed to clean only a limited area with a solvent-laden cloth and then discard it.

The contact adhesives are normally spread by brush or roller on both surfaces of the rubber sheet. The solvent is allowed to flash off; the contact adhesive becomes tacky to the touch. The sheets should not be spliced prematurely since solvents will be trapped within.

After the sheets are mated together, a hand-held roller is used to rigorously apply pressure across the lap joint. Again, a deliberate rolling technique should be followed. Workmen should not roll along the lap joint, stretching the sheet and causing a differential strain in the spliced sheets (top to bottom). The best technique is to go perpendicular to the lap joint in a to-and-fro motion to ensure intimate contact along the entire lap.

4.7.1.4   Hot Asphalt Method

The rubber-modified bitumen sheets normally call for a conventional roofing grade asphalt for making a lap joint. These sheets have a selvage edge and are part of a totally adhered roof membrane system. In fact, this system is very close to conventional BUR technology. The only deviation is the grade of asphalt and application temperature that may be specified by different manufacturers. An American Society for Testing and Materials (ASTM) D312 type III or type IV grade asphalt may be called for depending on the manufacturer's choice.

The rubber-modified bitumen membranes are rather sophisticated to manufacture when compared to conventional asphalt-saturated or asphalt-coated felts. Thus, the utilization of asphalt for the lap joint is a practical compromise since the asphalt is needed to adhere the sheets. It would be impractical to separately ship a special grade or mix of modified asphalt for lap splices only. Generally speaking, this method has given few field problems once the installation techniques and the behavior of the modified bitumen membrane has been well understood by the applicators.

4.7.1.5   Torch-Applied Method

Plastic-modified bitumen sheets use heat to adhere the sheet to an approved substrate, as well as to accomplish a cohesive lap splice to the adjoining sheet. It is normally done with open flame, propane-fueled torches. As a result, some contractors and building owners are concerned with fire safety. The torch-applied lap joints have normally been found to be quite adequate in strength and have demonstrated good watertight integrity.

The technique is quite simple. Direct flame is applied to the modified bitumen sheet until the material softens and begins to liquefy. The torch is then slowly moved along while the hot bitumen sheet is pressed together behind it. This technique also requires a degree of experience.

A good roof mechanic can torch lap joints together at a time-efficient rate. This type of lap joint is extremely versatile around roof penetrations and points where flashing details occur.

4.7.1.6   Self-Adhering Method

There are several varieties of self-adhering lap joints available. One previously discussed was a modified bitumen sheet that has a paper release strip. Other materials such as PIB, which is an uncured elastomer, have a butyl edge to make the lap joint. Again, a release paper is removed and a solvent wash is applied. The lap joint is then carefully rolled out to effect a watertight splice.

4.7.1.7   Tape-Splice Method

Tape splices may also be viewed as a self-adhering method for making lap joints. With rubber sheet membranes, tape splices are made by applying the self-adhering tape to the cleaned lap area. The overlapping sheet is then aligned and brought together with roller pressure applied. There are several different tape splices currently available, but they have not been widely used at this point.

4.7.2   Test Results

After lap joints are made, they should be able to exhibit a minimum strength (ideally equal to the membrane strength). Depending on the generic materials used for the lap joint and the type of single sheet itself, different amounts of ultimate elongation may also be of interest.

Tests should be performed at different temperatures in a manner that reflects the type of lap joint and its possible failure mode (Fig. 4-16). A straight shear test is normally conducted, which is a pulling apart of the lap joint in a straight, forward manner.

Another performance test is a peel test. A peel test may be run at 90° or 180°. It is similar to reaching down and pulling back on the edge of a lap splice. The test develops a force that creates high tensile stresses at the glue line or splice point of the lap. Figure 4-16 illustrates the lap shear and 180° peel test.

Shear and peel should be run under a variety of temperature and moisture conditions. Each manufacturer whose systems are not fully adhered should have shear and peel test results available.

4.7.3   Quality Assurance

Check all single-ply lap joints for integrity. Inspect every lineal foot. Normally, a steel blunt-nosed probe is used along the lap edge in an attempt to locate a false joint.

After this procedure, a lap sealant is then applied to the exposed edge. For

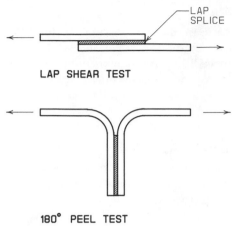

Figure 4-16. Lap-testing methods.

plain sheets, the lap sealant dresses off the edge and serves as a second line of defense. With torch-applied modified bitumen systems, a "buttered" edge, if properly done, provides a watertight edge. The lap joint itself is the primary barrier to water penetration.

Reinforced plastic or elastomeric roof sheets should have a lap sealant applied to prevent water from being wicked into the exposed reinforcement at the edge of the sheet. Studies have shown that water can be driven into this edge in varying amounts, depending on the temperature and style of reinforcement used. This effect was found in a variety of generic single-ply materials.

4.8   PREVENTIVE MAINTENANCE PROGRAM

If a PM program for single-ply roofs is to perform effectively it should consist of a number of well-structured and well-managed components. These components should include an efficient record-keeping system, effective inspection techniques for potential roofing failure detection, personnel well versed in the latest roofing maintenance methods, and cost accounting.

4.8.1   Record Keeping

Any PM program should include efficient record-keeping and documentation procedures. The filing system should provide for an easy access and updating of existing information and of documenting the repairs or maintenance activities. Include documentation of the facts of maintenance and repairs: materials used, dates repairs took place, and the extent of repairs made. Photographs may be highly beneficial if a change is suspected of taking place in the membrane. Include records of any roof-top equipment change or the installation of additional units. Keep the written documentation in one place and include a roof plan showing all penetrations and roof-mounted equipment. Detailed drawings of flashings and other termination points should be a part of this package.

Finally, include manufacturers' brochures and project specifications, as well as manufacturers' warranty documents in this file. (A copy of the manufacturers' warranties should also be located elsewhere for safe keeping.) The PM program is most effective when conducted on a regular basis with proper record-keeping procedures.

4.8.2   Procedures

A good PM program requires at least two roof inspections every year. Ideally, the first inspection should take place in the spring to check on damage that may have occurred in the winter. This inspection allows for repairs to be made during the fair weather months. The second inspection should be made in the fall after the prolonged exposure to heat and the sun to ensure that the roof is in good condition for the upcoming winter. In addition, the roof should be inspected following any major occurrence that may affect the roof such as severe storms, high winds, or rooftop construction activities. Checklists shown in Figures 3-12, 3-13, and 3-14 can be modified to deal with single-ply systems.

To locate moisture damage a number of nondestructive techniques can be employed, including infrared scanning, nuclear metering, and capacitance metering. Here are some considerations when choosing a nondestructive evaluation (NDE) method:

1. Select an appropriate system. Keep in mind that higher costs of equipment and skilled operators of some systems can be offset by the system's efficiency making the particular survey less expensive in the long run.
2. All systems require dry surfaces.
3. Infrared scans must be done at night to get accurate results.
4. Core samples are needed to clarify results and avoid false readings. Both wet and dry areas should be checked.
5. Always check the roofing material manufacturer's guarantee or warranty before taking a core sample to assure requirement compliance.
6. Make sure core samples are patched by a qualified installer.

7. Insist on a visual inspection of the roof, in addition to the NDE and core samples.
8. If a survey has been done, make sure it was done correctly by qualified individuals who have a list of clients and references.

For a comparison of different methods on NDE of moisture in roofing systems see Table 4-1. Prior to making a detailed inspection of the roof, make

Table 4-1.   Determining Which Nondestructive Roof Moisture Evaluation Method Should Be Used

	Evaluation Methods		
	Capacitance	Nuclear	Infrared
Conventional BURs			
BUR (no insulation)	‡	−	−
Insulation type			
Fiberboard	‡	‡	‡
Perlite board	‡	‡	‡
Fiberglas	‡	‡	‡
Urethane/Perlite composite	‡	‡	‡
Urethane	‡	‡	‡
Extruded polystyrene		−	−
Beadboard (expanded polystyrene)	‡	‡	‡
Cellular glass (foam glass)	‡	−	−
Lightweight concrete			
All-weathercrete	‡	−	
Urethane with foil backing	−	‡	‡
Sprayed-on urethane			‡
Single-Ply Membrane Systems			
(with commonly used insulations,			
i.e., EPS, urethane)			
Neoprene, ballasted	+	+	−
Neoprene, unballasted	+	+	+
PVC, ballasted	+	+	−
PVC, unballasted	+	+	+
CPE, ballasted	+	+	−
CPE, unballasted	+	+	+
CSPE, ballasted	+	+	−
CSPE, unballasted	+	+	+
PIB, ballasted	+	+	−
PIB, unballasted	+	+	+
EPDM, ballasted	−	+	−
EPDM, unballasted	−	+	+
Inverted roof membrane assembly system	−	−	−

Notes: + represents good value; = represents limited value; − represents no value.

a visual inspection of the structure below. The purpose of this inspection is to detect any water staining, rusting, or spalling of the deck, and any problems related to drain leaders. Also, look for any other indications of possible roofing problems.

Any changes of occupancy of the building that could create different environmental influences for the roof system should also be examined. This case is especially true in process industries where chemicals, machinery vibration, or interior humidity and temperatures may change significantly from time to time. Check out any indication of leaking or weather staining on the ceiling, which could be caused by condensation of uninsulated piping or duct work and may not be related to roof failures. Similarly, check exterior walls for cracked or open mortar joints, spalling efflorescence, or leaking copings.

Water staining or rusting of steel decks or water staining of the ceiling, if not caused by condensation or by water entry through mechanical equipment, is undoubtedly caused by problems in the roofing system. The causes of water entry into the building must be found before any maintenance work or minor repair work commences. To uncover the source of problems, the rooftop surface, flashings, drains, pitch pans, and hardware must be examined.

When dealing with the roof surface, look for signs that indicate a possible failure of the roofing membrane. The inspector should examine the possibility of open lap joints, a serious problem on a single-ply roofing system (Fig. 4-17). Unfortunately, open joints are not that easy to locate. Since they cannot be readily seen, about the only way to find open lap joints is to check the joint by hand. Of course, once open joints are located, they should be resealed immediately.

Fish mouthing is another type of joint failure (Fig. 4-18), although it does not occur frequently in single-ply membranes. Fish mouths should be cut out and stripped in if they are found. Ridging also needs to be looked for, although this type of defect generally does not cause serious problems for single-ply systems.

In many instances, unadhered areas on fully adhered systems could be responsible for serious failures. Flopping of these unadhered sections in high winds could loosen larger areas of membranes, making them susceptible to tearing. Locating unadhered areas can be somewhat difficult. Many times suction cups have been employed to find these defects. Once these unadhered areas have been located, they can be mechanically fastened or cut open and re-adhered. Any cuts or mechanical fasteners, of course, would have to be stripped in to ensure that water does not enter into the building through the fastener or the cut.

Another potential failure is fastener back-out, which happens on mechanically fastened systems. The protruding fastener makes the membrane highly susceptible to puncture, especially, if it is subjected to any mechanical impact.

Figure 4-17. This roof contains wet insulation resulting from water entry through the lap. When stepping on this area of the roof, the water rushes from the roof insulation onto the roof surface.

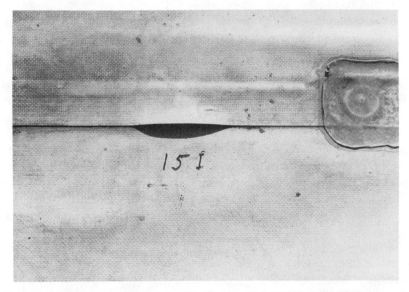

Figure 4-18. Fish mouths are an indication of an inappropriately constructed lap.

Backed-out fasteners should be removed, new fasteners installed, and the area stripped in.

When inspecting a roof, nonuniform depth of the stone on a ballasted system may be discovered. It is usually the result of wind forces on the roof surface. Although it is not necessarily a serious problem, the ballast should be redistributed to make sure that membranes are adequately held down over the entire roof.

Many of the problems of single-ply roofs are caused by ponded water and poor drainage. Poor drainage is caused by an inadequate slope of the roof, deflection between vertical supports, or obstructed drainage. Areas that tend to pond can be corrected by installing additional drains. If the drains are plugged, the debris that causes the problem should be removed. The removal of debris should be a routine PM activity.

If cracking, generally a surface failure, of single-ply membranes is observed, there is most likely a problem with the membrane material itself and the manufacturer should be notified. Splitting, a defect that extends throughout the membrane, can occur when wind causes unadhered sheets to flap and fatigue in extreme cases. If sections of the membrane in a ballasted system are loose, they should be weighted down with an appropriate amount of ballast. If the system is a mechanically fastened or fully adhered one, it should be mechanically refastened or re-adhered. When inspecting the roof, also look for discoloration and changes in the membrane's surface texture that could be due to changing chemical properties in the membrane. If this situation is occurring, the membrane has to be carefully monitored. These defects can also be caused by the materials' incompatibility in the makeup of the membrane or from weathering or chemical attack. Chemicals could be present in the atmosphere or could be produced by the processes taking place within the building.

Base and counterflashings require careful attention in the single-ply roofing system (Fig. 4-19). Flashing problems can result from failed caulking, in which case deteriorating caulking should be removed and the joint should be recaulked. Flashing failure also occurs as a result of fastener failures. Backed-out fasteners cause loss of attachment of flashings. Survey all flashing and termination bar assemblies to ensure that caulking and fasteners are in place and functioning.

With metal flashings, corrosion could be a problem. In many instances, recoating of metal flashings is all that is necessary to ensure their long-term performance. Pitch pans are components of a roofing system that are prone to failures. If entrapped air and networks of "caves" are seen in pitch pans, they need repair. Inspect all pitch pans to make sure they are filled and properly sealed. Add sealer as necessary, ensuring that the new sealer is compatible with the material that is already in the pitch pans.

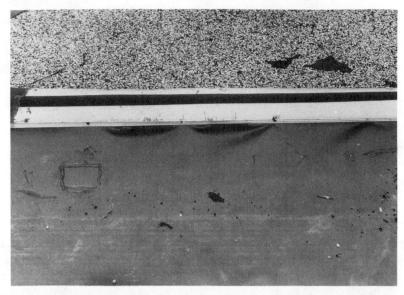

Figure 4-19. Inadequate coverage of base flashing by counterflashing.

Many people are surprised to see the enormous amount of mechanical equipment on the roof. This equipment includes exhaust fans, rooftop air-conditioning units, and piping. Often, much of this equipment is improperly supported, due to faulty initial design and construction. Also, mechanical equipment requires maintenance, which causes significant traffic and moving of heavy equipment. Inspect all hardware and metal cabinets, fan shrouds, and hoods to ensure attachment. Refasten any loose equipment. Repeated wind forces may cause a piece of the metal cabinet to break loose, which may cause cuts and abrasions on mechanically fastened or fully adhered single-ply membranes. (Ballasted membranes are not as susceptible to the damage caused by equipment on the roof.)

It is imperative that roof traffic be controlled by confining it to certain areas that are protected against foot traffic and equipment moving. After any significant repairs of mechanical equipment on the roof, inspect the roof and repair any damage without delay.

When performing PM work, the principles of new construction may be applied. One main difference is that in PM work materials have been exposed to weather, sometimes for many years. Therefore, special precautions must be taken to ensure the surfaces are adequately prepared for receiving new material. Compatibility of the material on the roof and new repair material should be of foremost concern.

4.8.3 Program Costs

Single-ply roof systems can cover wide-open uncluttered roofs such as warehouses and shipping facilities or be used to roof a process building that has numerous stacks, penetrations, and pitch pockets. Depending on the roof system and complexity of design, a thorough rooftop inspection may require one man-hour per 5,000 ft^2. Large, expansion roofs may require one man-hour per 25,000–30,000 ft^2. Documentation of findings will take a substantial amount of office time and may result in an additional 25% of the total time spent on the actual rooftop inspection. If repairs are needed, additional time will be required, either from in-house maintenance or a qualified roofing contractor.

4.9 SUMMARY

Roofs are constantly under attack by the forces of weathering, building stresses, and mechanical and/or chemical action, either from rooftop traffic or the atmosphere. All roofs undergo normal aging, but major problems arise from neglect, abuse, and contamination. The intent of PM is to avoid costly repairs by preventing premature failure of the roofing system.

A good PM program can only come about if the owner, designer, and contractor are knowledgeable about the materials, design, and construction techniques employed for roofing systems. For early detection of roofing problems, a thorough semiannual inspection is critical. If the possibility for problems appears, the inspection will, at a minimum, alert the owner that more serious investigation is required.

By preventing catastrophic roof failures, PM also pays dividends through minimizing the interruption of activities taking place within the building. Most important, PM can protect the investment made in the roofing system.

4.10 SUGGESTED READINGS

Baker, Maxwell C. 1980. *Roofs: Design, Application and Maintenance*. Montreal, Quebec, Canada: Multiscience Publications Ltd.

Dupuis, Rene. 1989. How to prepare comprehensive reroofing specifications. In *Handbook of Commercial Roofing Systems*, pp. 6–14. Cleveland, Ohio: Edgell Communications, Inc.

1983 Handbook of Single-Ply Roofing Systems. New York: Harcourt Brace Jovanovich Publications.

Henshell, Justin, and Paul Buccellato. 1989. Detailing flashings on single-ply roofing systems. *Construction Specifier* **42:**(11):72–75.

Lewis, H. Z. 1983. Moisture detection in plant roofing systems. *Plant Engineering* **37**:92–93.

The NRCA Roofing and Waterproofing Manual, 3rd ed. 1989. Chicago: National Roofing Contractors' Association.

A Professional's Guide to Single-Ply Roofing Specification. 1984. Glenview IL: Single Ply Roofing Institute.

Tobiasson, Wayne. 1982. Proceedings of the 1982 International Society of Optical Engineers meeting. *Roof Moisture Surveys: Current State of Technology.* Bellingham, WA: Photo-Optical Instrumentation Engineers.

Tobiasson, Wayne, and Charles Korhonen. 1985. Proceedings of the 1985 International Symposium on Roofing Technology. *Roof Moisture Surveys: Yesterday, Today and Tomorrow.* Chicago: National Roofing Contractors' Association.

4.11   ASSOCIATIONS WITH INFORMATION ON SINGLE-PLY ROOFING SYSTEMS

American Society for Testing and Materials (ASTM)
1916 Race Street
Philadelphia, PA 19103

Asphalt Roofing Manufacturers Association (ARMA)
6288 Montrose Road
Rockville, MD 20852

Cold Regions Research and Engineering Laboratory (CRREL)
72 Lyme Road
Hanover, NY 03755

Factory Mutual System (FM)
1151 Boston-Providence Turnpike
Norwood, MA 02063

Institute of Roofing and Waterproofing Consultants
51 West Seegers Road
Arlington Heights, IL 60005

Midwest Roofing Contractors Association (MRCA)
8725 Rosehill Road, Suite 210
Lenexa, KS 66215-4611

National Roofing Contractors' Association (NRCA)
6250 River Road
Rosemont, IL 60018

The Roofing Industry Educational Institute (RIEI)
14 Inverness Drive East
Building H, Suite 110
Englewood, CO 80112-5608

The Single-Ply Roofing Institute (SPRI)
1800 Pickwick Avenue
Glenview, IL 60025

Steel Deck Institute (SDI)
P.O. Box 3812
St. Louis, MO 63122

Underwriters Laboratories (UL), Inc.
333 Pfingsten Road
Northbrook, IL 60062

5. Exposed Metal Roof Systems

Len E. Lewandowski

5.1 INTRODUCTION

Exposed metal roofs are currently being installed on 57% of all new, low-rise, nonresidential buildings. Since 1970, the use of exposed metal roof systems has expanded from exclusive use on metal buildings to all types of conventional and specialty buildings because of advancing technology in cold-formed metal fabrication. In addition, the relatively new concept of retrofitting older, conventionally roofed buildings with new, high-technology, metal systems has emerged. Industry sources reported that over 1.3 billion ft^2 of structural, metal roofing was commissioned in 1988.

The old-style "tin roof" has long since given way to systems that base themselves on architectural adaptability, durability, and low maintenance. Further, life cycle costing studies have shown metal roof systems to be competitive with any roofing system currently on the market. When considering the variety of designs, materials, and finishes of metal roofs being offered today, the options can become overwhelming. Nearly every available metal roof system, however, can simply be classified as either architectural or structural.

5.2 ARCHITECTURAL METAL ROOFS

An architectural metal roof is basically a decorative, surface treatment. It offers a degree of protection from the weather, but it is not considered a structurally active element. This type of roof is usually installed over a structurally supported wood (plywood sheathing) or steel (metal decking) substrate that has, in turn, been topped with a vapor retarder, insulation board, and moisture-resistant roofing felts (Fig. 5-1). A minimum roof slope of 3 in./ft of run is typically recommended to ensure water shedding of the

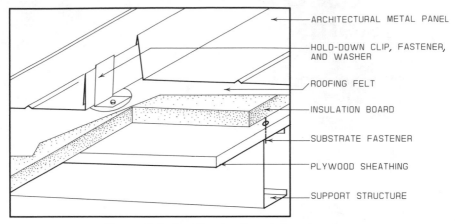

ARCHITECTURAL METAL PANEL

HOLD-DOWN CLIP, FASTENER, AND WASHER

ROOFING FELT

INSULATION BOARD

SUBSTRATE FASTENER

PLYWOOD SHEATHING

SUPPORT STRUCTURE

Figure 5-1. Architectural metal roof system.

BATTEN SEAM
W / CLIP

SNAP SEAM
W / CLIP

INTERLOCK SEAM
W/O CLIP

Figure 5-2. Clips used for anchoring architectural metal roof panels.

metal membrane surface. Even steeper slopes are often specified to maximize the visual exposure of the surface for its aesthetic value.

Architectural metal roofs may be fabricated from steel, aluminum, zinc, copper, lead, stainless steel, or a composite alloy material. The panels are usually continuous (no horizontal end laps) over relatively short slope runs (typically 40 ft maximum length). They can be anchored to the wood or metal substrate by direct fastening (through-fastening) or by concealed hold-down clips (Fig. 5-2). Side seams may be interlocked, snapped together, or batten-sealed and may or may not incorporate a sealant material.

Since the application of an architectural metal roof is essentially decorative, maintenance is primarily concerned with the care of the surface finish. Given that appearance is of overriding concern, and roof slopes are relatively steep, extreme care must be exercised during the implementation of any prescribed maintenance procedure.

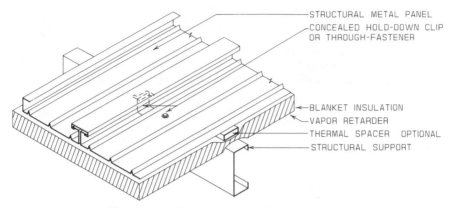

STRUCTURAL METAL PANEL
CONCEALED HOLD-DOWN CLIP
OR THROUGH-FASTENER

BLANKET INSULATION
VAPOR RETARDER
THERMAL SPACER   OPTIONAL
STRUCTURAL SUPPORT

Figure 5-3. Structural metal roof system.

5.3   STRUCTURAL METAL ROOFS

The structural metal roof is a multifunctional system that serves a variety of purposes. Weather protection is, of course, its paramount and most obvious function. But it also embodies a host of almost equally desirable, yet less apparent, characteristics. Load transmission and support element stabilization are two functions that are extremely valuable from an engineering standpoint. In some system types ("stress-skin" assemblies), the roof is actually considered to be a primary element of the superstructural assembly. More commonly, however, a structural metal roof consists of a system of independent, synergistically integrated components assembled to provide the form and function of a building roof (Fig. 5-3).

The main difference between architectural and structural metal roofs is the incorporation of a substrate. Structural metal roofs do not require an underlying substrate for the provision of strength or for moisture protection. The metal panel itself acts as weather barrier and load distributor. Roof live loads and, in some cases, lateral loads are transmitted directly to the spanning structural members by the metal membrane.

Structural metal panels are typically fabricated from coiled carbon sheet steel that is galvanized, aluminized, or aluminum-zinc-coated by a hot-dip method. Zinc coatings provide sacrificial (self-healing) corrosion protection. Aluminum coatings are primarily a barrier protection. Zinc and aluminum composites offer limited degrees of both. The sheet may also be finish-painted with a polyester, siliconized polyester, or fluorocarbon coating, or laminated with an adhesively bonded acrylic or fluorocarbon plastic film. Paint and laminate films are applied to enhance appearance and increase barrier protection for corrosion resistance. The coated and recoiled material is later roll-formed into its final profile and cut to length.

Structural metal roofs are categorized into two groups based on the method used to anchor the panels to the spanning substructure. Attachment can be made by either exposed, through-membrane fastening (as through-fastened or screwed-down roofs) or by concealed anchor clips (as standing seam roofs). The side seam and end lap joints of both types contain either factory- or field-applied sealants. It is the judicious use of joint sealants that renders the membrane moisture-proof. As a result of their water tightness and strength capabilities, structural metal roofs can safely and effectively accommodate slopes down to ¼ in./ft of run.

5.3.1   Through-Fastened Metal Roofs

The through-fastened or screwed-down metal roof is a proven system that has been used on low- to midrise nonresidential construction ("metal buildings") for decades (Fig. 5-4). The typical panel is roll-formed into its finished corrugated profile from a mild carbon sheet steel that is hot-dip galvanized. It may also be finish-painted or laminated. Adjoining sheets are commonly interlapped. Fastening at the end laps is made through both sheets and directly into the spanning substructural component (usually a light gage purlin). At the side laps (generally 1½ in. above the drainage plane), both sheets are stitched together with sheet metal screws. Fasteners are typically hex-washer-head, self-drilling, and/or self-tapping sheet metal screws with integral metal-neoprene washers. Foam or rubber closures and tape- and gun-grade sealant materials are used at eave, side lap, end lap, ridge, and rake conditions. Closures and sealants are usually field-applied.

This roof system is recommended for installation on roof slopes of ½ in./ft of run or greater. The length of a continuous slope run is limited only by the flexibility of the underlying support structure. One of the drawbacks of a

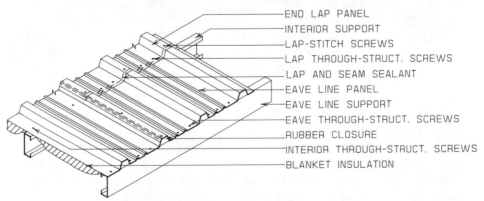

Figure 5-4. Through-fastened (screwed-down) metal roof.

screwed-down system is its high susceptibility to thermal shock stress. Shock stresses develop as a result of the membrane's positive attachment to and the restraint supplied by the supporting substructural components. For this reason, a screwed-down exposed system is not recommended for use over stiff substructures, such as joists or plate sections. The relative stiffnesses of the roof and the support system should be factored into the determination of the maximum permissible surface dimension between expansion/contraction relief joints. The recommended practical limit is 200 lineal feet.

Through-fastened roof systems are one of the lowest initial cost, structural roof assemblies available. The drawback of positive mechanical attachment turns into a definitive advantage when considering that this roof has the ability to absorb and redistribute lateral forces to the building bracing. It is the through-membrane fastening of the roof sheet to the substructure that gives the roof this diaphragmatic capacity. In typical applications, diaphragm capacity will reduce the amount of required bracing in the building, which will translate into a cost savings in both material and erection. Other major advantages of a screwed-down system include its lightweight design, installation simplicity, and low maintenance requirements.

5.3.2   Standing Seam Metal Roofs

The standing seam roof (SSR) is the most recent advancement in structural metal roofing technology (Fig. 5-5). This roof design, when properly constructed, can virtually eliminate the leak potential associated with through-fastener penetrations in the drainage plane of the membrane surface. It employs a

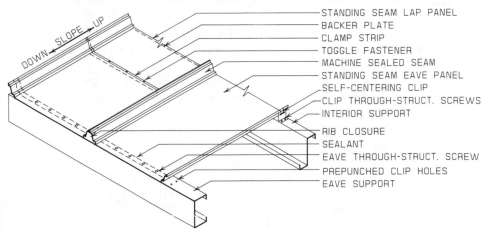

Figure 5-5. Standing seam roof system.

sophisticated, concealed, clip mechanism to secure the roof sheets to the building substructure. Side seam joints are battened, snap-locked, or machine-rolled and "stand" 2 or more inches above the roof's drainage surface.

SSR sheets are typically roll-formed from high-tensile carbon sheet steels that are aluminized or aluminum-zinc-coated. Painted or laminated film finishes are also optionally available. Side seam joints commonly incorporate a factory-installed sealant to provide weather tightness. End lap joints are achieved by stitching the sheet ends together with either sheet metal screws or toggle-type fasteners. Some systems also employ clamp strips and backer plates to increase joint stiffness and provide a tighter seal. A variety of metal and neoprene closures, sealant materials, and sheet metal fasteners are used to weatherproof the eave, ridge, and rake lines, and other surface break or transition conditions.

An SSR can be installed on roof slopes as low as ¼ in./ft of run. Some manufacturers offer the option of a "fixed" anchorage SSR for smaller-area roof surfaces. Expansion and contraction force development is kept at a low level by limiting the expanse of the continuous membrane. A fixed SSR is similar to a through-fastened roof with respect to its sensitivity to the relative stiffness of its underlying support system. The elevated seam is promoted for its leak resistance in fixed systems.

Most SSR installations are the "floating" (laterally independent of the support structure) type. In this case, the maximum length of uninterrupted slope run is dependent on the degree of expansion/contraction freedom afforded by an anchorage clip mechanism. The common range is 100–200 feet. The advantage a mechanized clip anchor provides is that it effectively makes the roof surface "float" above the support structure. This independence eliminates the possibility of high thermal shock stress development within the system. The absence of this distress mode significantly reduces or eliminates the need for potentially troublesome expansion/contraction relief joints in the roof surface. As a result, SSR's can be used over stiff subsystems. They also offer greater latitude with respect to the incorporation of high-efficiency insulation systems.

The concept of membrane independence (the floating roof) is by no means a panacea. It does impose some disadvantageous conditions upon the design. In this floating state, the roof surface has no lateral load resistance capability. It is nondiaphragmatic. Typically it can neither absorb nor transmit horizontal forces. Therefore, the lateral load and stability forces acting upon and within the building must be accommodated by other means. As a result, the design and installation of an SSR becomes more complex than a screwed-down roof system. There are relatively higher material and installation costs. Even with its increased complexity, the standing seam roof offers the lightest weight, lowest maintenance, and most cost-effective roofing solution available today.

5.4 DESIGN CONSIDERATIONS

Good design is the basis for minimizing the PM and repair of any building system. In consideration of all the possible factors that may contribute to the premature failure of structural metal roofs, a valid question to pose is, What keeps them from failing? Even though it is not fool-proof, a structural metal roof can be an extremely attractive and effective investment if it is designed and installed properly. There are, in fact, a number of judicious considerations that must be taken into account during the design phase of any new construction project. The inclusion of some specific ones regarding metal roofs can significantly reduce the probability of a premature failure. Most of these considerations amount to little more than common sense.

Each mechanical component employed in a structural metal roof system is factory-mass-produced to stringent quality standards. The very same fabrication standards and controls that ensure product excellence limit that product's application flexibility as well. By complicating the roof design with unique architectural corners, oblique angles, and multiple transitions, the installer is forced to employ his own resourcefulness in compromising the limits imposed by factory standardization. Most times this approach will result in a measurable degree of compromised performance.

In addition to typically standard design considerations regarding code compliance, load capacities, and insurance ramifications, a metal roof system has some unique characteristics that merit careful deliberation. Since a metal roof system must perform the multiple functions of a structural component as well as a protective covering, careful consideration must be given to its compatibility with the underlying support structure (type and module), eave, ridge, and rake line junctures and any accessory integrations that are incorporated into or penetrate the membrane.

Because an SSR system is a coagulation of diverse yet interdependent components, most of these systems are offered only as all-inclusive installations or as part of a total building package. Each item may be specifically designed for use with another specifically designed companion element. This situation severely limits the number of "or equal" options available to the designer. The colloquialism "parts is parts" does not apply to metal roof systems. The choice of a structural metal roof must be completely integrated into the building's overall design concept.

5.4.1 Roof Movement

A principal design consideration is the roof's expected movement. What is the projected extent of temperature and barometric pressure variation to which it will be subjected? Would it be advantageous or detrimental to develop a diaphragmatic capability in the roof membrane (screwed-down roof systems)? Generally, roof diaphragms are advantageous on smaller, single-story, cubular structures (100 × 100 ft) where the only lateral factors are wind and stability. It is recommended that taller buildings, structures with high length-to-width ratios, earthquake-sensitive constructions, and buildings subject to auxiliary load vibrations (cranes, conveyors) have a bracing system that functions independently of the roof diaphragm. In these instances, a floating SSR may be the wiser choice since it requires no "balancing" of diaphragmatic and independent bracing system stiffnesses. Caution must be exercised by the designer when specifying a system whose performance is dependent on stiffness balancing. If the roof diaphragm is allowed to be effectively stiffer than the independent bracing system, lateral forces will be absorbed by the roof first. The result may be severe joint fatigue and an imminent performance failure within the roof membrane.

5.4.2 Roof Slope

Roof slope is another major consideration. A roof's life expectancy is heavily contingent on efficient removal of moisture from the metal membrane surface. The most effective way to prevent moisture penetration is to quickly shed the water from the roof over as short a path as possible. Minimum recommended slopes are ½ in./ft of run for screwed-down roofs and ¼ in./ft for standing seam systems. Free drainage planes should be unimpeded. Length of slope run and potential wind and accessory penetration damming are important contributing factors. Can the roof system be fitted with flow diverters to channel the water around large penetrations (Fig. 5-6)?

Figure 5-6. Flow diverters channel water around large penetrations.

Drainage system capacity (gutter size, length and number of drops) should be carefully analyzed to prevent ponding and overflow situations from developing. Valley gutter conditions should be avoided. Analyze the impact that maximizing the roof slope has on material cost, HVAC (heating, ventilation, air conditioning) requirements, and building aesthetics.

5.4.3   Roof Details

Because metal roof systems have demonstrated themselves to be extremely installation-sensitive, designers have learned to look for the inclusion of certain proprietary details in the systems they specify for their projects.

Roof sheet, end lap conditions should be factory-fabricated. All required eave, rake, and ridge condition materials should be included as part of a total roof system installation package. Each flashing and trim component should be specifically designed for integration into the selected roof.

The most important aspect is transition joint compatibility. If the roof panels are anchored with a device that facilitates expansion and contraction, but are positively attached to flashing and trim materials that do not permit movement, a condition of differential restraint exists. Something must give when an element wants to move relative to one whose position is fixed. It will usually be the joint seal. All thermal movements must be accommodated. Compound joints (joints that absorb multidirectional thermal movements) must be provided where compound thermal forces are expected such as rake line/ridge line intersections (Fig. 5-7).

Look for a system that limits the number of through-fastener penetrations to an absolute minimum. If considering an SSR, look for a two-piece clip

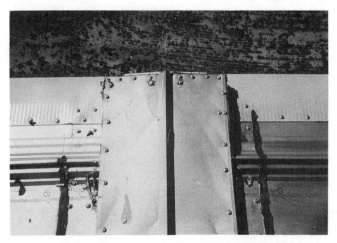

Figure 5-7. An ineffective compount joint.

mechanism with a self-centering expansion/contraction tab. The self-centering feature keeps the tab in a neutral position regardless of sheet temperature at the time of installation, which ensures that tab runout and resultant clip binding will not become a problem with extreme temperature variations. Also check the relative positions of the interlocked clip and the factory sealant. Thermal movements should not effect the seam sealant in any way.

All field-installed rubber closure and sealant materials should be specifically tailored for use with the selected roof system. Rubber closures should be form-fitted and chemically compatible with contact surface finishes. Sealants must also be chemically compatible with contact surfaces, have good flow characteristics, and maintain their resiliency. Sealants containing acetic acid must not be used. Both closures and sealants must be resistant to degradation from sun, moisture, ozone, and marine or industrial contaminants.

5.4.4   Rooftop Penetrations

Rooftop penetrations are second only to fasteners in receiving blame for a leakage failure. In many instances where the penetration is properly identified as the culprit, the original roof installer may be blameless since it is common for roof-penetrating equipment to be installed well after the roof itself is completed. The sealing of the membrane around the protrusions may have been left to tradesmen who lack the proper skills, equipment, and materials necessary to effectuate a weathertight joint. A penetration is simply a stoppered hole in the roof that is bordered by a continuous joint. This joint must meet all the requirements of an eave, rake, and ridge joint and more. Transition flashings, trim, and sealants must be compatibly integrated with the mechanics of the roof membrane. Do not compromise the integrity of this critical location with inexperience. Insist that all roof modifications be made by a qualified installer.

Roof penetrations should be kept to an absolute minimum. Vent through the wall if at all possible. Group those that must go through the roof and keep them independent of the membrane surface. If a unit must be structurally supported, use an above-deck frame with round pipe columns that are directly supported by the building substructure. Avoid using generic curbs unless the unit is light enough to be supported directly by the deck and small enough to fit between the lines of substructure. When specifying above-deck accessory framing, be sure to include a protective coating that provides adequate protection against runoff staining.

5.4.5   Designer and Installer Qualifications

Finally, the complexity and propriety of SSR roof systems demand the selection of a designer and installer familiar with and experienced in the selected system. Ask for references. Visit completed installations. Walk the roof with the maintenance manager if possible and ask for his opinion. Check out the fabricator's organization as well. Does he have dedicated field service personnel or is the salesperson also the engineering, materials, and installation expert? Is full-time factory supervision available for the installation phase? Do not rely solely on a manufacturer's material warranty. Most provide only limited protection against deterioration due to material defect. However, reading the warranty will explain what the roof is not protected against. Most reputable contractors offer building performance warranties that include leak coverage for a limited period. Make sure this protection extends to failure due to manufacturing and design defect as well as installation-induced problems.

5.5 PERFORMANCE FAILURES

A structural metal roof can represent upward of 70% of the exposed building surface area. Its natural enemies are water, wind, sun, and temperature fluctuations. In many instances, these natural enemies are allied with ignorance, incompetence, or neglect. It has been estimated that over 70% of the litigation brought against the building construction industry is the result of some sort of roof performance failure. The failure is usually a leak, or a condition that will ultimately result in a leak. Suddenly, high expectations of "superior maintenance-free performance" may turn into bitter and even hostile expressions of disappointment.

Metal roof performance failures are likely to be traceable to one of the following factors:

1. Manufacturing error
2. Design error
3. Installation error
4. Owner abuse (including neglect of normal periodic maintenance)
5. Combinations of some of the above
6. Combinations of all of the above

Many investigative professionals report that a performance failure within the first 2 years of a roof's life is likely to be installation-related. Barring obvious owner abuse, those failures that develop after 2 or more years of satisfactory service usually signal an underlying material, system, or design problem. Most failures are caused by combinations of these factors. In an attempt to provide a measure of consistency regarding the classification of performance failures as they relate to probable causation, it is beneficial to establish a set of basic definitions for each error factor.

Manufacturing errors are defects or inconsistencies in the material, material specifications, and/or the material installation instructions of the roof system supplier. Examples of manufacturing errors are cracks in the radial bends of the formed panel, accelerated corrosion as a result of poor or incompatible factory-supplied material finishes, or installation instructions that result in differentially restraining the expansion/contraction movement of an SSR.

Design errors are inconsistencies within the material specification or material utilization details established by the architect, engineer, or designer of record. Examples of a design error include attachment or a through-fastened roof directly into a steel joist substructure, failure to specify a slope that is adequate to promote free drainage, or specification of roof materials or finishes that are not compatible with environmental conditions.

Installation errors are defects or deviations with respect to recommended materials, procedures, and sequences of erection, as they relate to the preservation of the design concept integrity, due to ignorance or negligence on the part of the installer. Examples are positively fastening a floating roof system directly to the spanning substructure; failure to follow the manufacturer's recommendations for sheet installation or placement of fasteners, closures, or sealants at joint conditions; and creating localized restraint in a floating system by employing improper penetration techniques.

Owner abuse is an action initiated by or at the direction of a building owner or occupant that is contrary to manufacturer care recommendations and effectively undermines the original roof system design concept. Examples of abuse would be damage to roof system components or finishes as a result of owner-installed rooftop accessories or equipment and failure to provide normal periodic maintenance.

As stated earlier, it is common for a performance failure to result from a combination of one or more of the above errors. Errors may be further compounded by the trauma that sets in upon discovery of a leak. A panic develops among the concerned parties, and haphazard fixes are randomly prescribed and, unfortunately, implemented. Many times they result in worsening or recurring leak problems. A small amount of insight as to where to look for a roof leak, and why the leak occurs, may save unnecessary expenditures of time and money. It will also aid in the determination of the most effective remedial course to follow.

5.6   JOINTS: THE KEY FAILURE AREA

Roof joints have been described as leaks waiting to happen. Ninety-nine percent of all leaks will originate at a joint. Only a perforation through the membrane exceeds the leak potential of a joint. Joints are located at eave, rake, ridge, valley, elevation break, expansion, and hip lines. They also occur at end laps, side laps, and surface penetrations. The design conditions of each joint vary, and the installation factors affecting performance compound these variables, which makes the isolation of the cause of a performance failure difficult.

The proper fit-up and sealing of the components that constitute a joint assembly is vital to overall roof system performance. The recommended details and sequences of assembly must be precisely followed to ensure maintenance of the design integrity. Each joint element plays a crucial role in achieving a leak-proof installation. There is no contingency for error.

Unfortunately, many roofs will occasionally provide an opportunity to determine the cause of a leak and to make a subsequent evaluation of the effectiveness of a remedial leak prescription. All elements of a suspect leak location must be meticulously analyzed in a failure investigation. It may sound like a lot of trouble, but it is necessary.

5.7   INSPECTION PROCEDURES

Many designers do not feel that the appearance of a structural metal roof system adequately complements their architectural statement. They reduce or eliminate casual visual exposure by specifying low slopes and sight barriers such as perimeter facades. In addition, a catch phrase in the metal roofing industry is "maintenance-free." All too commonly, the result is an out-of-sight, out-of-mind syndrome, steeped in a false sense of indestructibility. This cavalier neglect usually lasts until the first serious performance problem develops, and the owner discovers that maintenance is anything but free.

An examination of relevant historical data would seem to indicate that the maintenance-free roof has not yet been developed. Maintenance requirements are, at best, periodic and, at worse, litigated. All roofs, metal or otherwise, should be inspected on a regular basis.

The first inspection should be made with the contractor and designer immediately upon completion of the building. It gives the owner an immediate basis for comparison. Does this new installation look like the role model on which the original choice was based? Is the workmanship tight, straight, and professional especially at the joints? Has the contractor performed his housekeeping properly? No loose material should be left on the roof, including metal trimmings, tools, extra material, metal shavings, metal particles, or mud. Every panel should have a wiped clean appearance. Do not forget to check the gutters as well. All handling damage should be repaired to the point of being unnoticeable. Finish integrity should be properly restored in all mechanically or chemically distressed areas. It is much easier to get these things rectified before, rather than after, the contractor leaves the site. If everything is in order and the building does not develop immediate leaks, the owner has a clear picture of what a sound, weathertight installation should look like. Photographs of critical areas will preserve this impression for future comparison. After this initial acceptance inspection, regularly scheduled check-off inspections should be incorporated into an overall building maintenance program. Check-off inspections should take place every 6 months. Supplemental inspections should also be made immediately after a severe storm or any significant rooftop activity such as the installation of mechanical apparatus.

A roof inspection, activity, and maintenance file should be established. This file should contain *all* documentation pertaining to the construction and maintenance of the roof system. It should include plans, specifications, installation instructions, user manuals, warranties, and any other data that may be reasonably associated with current or future roof performance. All roof inspection reports should be kept here. Every roof activity should be logged into the file. Each activity should be systematically documented including information as to the date, time, type, and extent of the activity and the roof area accessed or affected. Keep the file current. It is much less costly to look up precise information than it is to reconstruct it if it is lost or missing.

A maintenance policy, even if informal, should include stipulations such as never allowing unauthorized personnel access to the roof; keeping all necessary, authorized traffic to an absolute minimum (safety considerations, however, dictate that no one should be on a roof alone); and permitting only qualified roofing experts to make any future modifications to the membrane surface, especially the cutting in of penetrations for accessories or equipment. Even minor repairs should be left to an expert.

Many owners prefer to contract professionals to perform periodic follow-up roof inspections. Professional inspections usually include a detailed examination of existing conditions followed by a comprehensive written report. The report will generally contain a description of the current condition of the roof as it relates to its life expectancy. Certain specific conditions may be cited with regard to leak potential. This report will normally be accompanied by a list of recommendations for the maintenance or improvement of the stated conditions as well as a cost estimate for their implementation. The owner can then make a knowledgeable business decision based on a cost–benefit evaluation. Many professional inspection firms offer full-service routine evaluation and repair packages that are backed by watertight guarantees.

The cost of inspection services can vary widely. A complete, detailed, routine inspection of an average-size (25,000 ft^2) roof should take no more than a day. An experienced installer may charge $300–$500 for this inspection service. The cost includes a written report and may also include a few minor repairs if they can be performed during the course of the inspection with minimal materials. If required repairs are more extensive, the professional will typically submit repair cost estimates along with the report. This expense may actually be viewed as a bargain when considering that the initial investment in the metal membrane alone for a 25,000 ft^2 surface is likely to be in excess of $50,000. If major roof problems are allowed to develop, an engineer experienced in these systems is likely to charge up to $5,000 for an in-depth structural evaluation that might tender only one recommendation: total replacement. The cost of replacing a 25,000 ft² roof could exceed $250,000. In the face of this risk, the value of an organized roof maintenance program becomes obvious.

At the very least, the building owner should perform a spring and fall walkover of the roof. It must be understood that this exercise is intended to supplement a committed program for roof inspection, not substitute for it. As is the case with any technically oriented product, it may take a trained technician to recognize a potentially troublesome and costly situation. A few key conditions can be owner-evaluated and possibly corrected. These include the following:

Roof Debris: All debris should be swept up, picked up, bagged, and removed. If a significant deposition of airborne contaminants is noticeable, it is also a good idea to hose or mop down the surface.

Drainage: Inspect roof gutters and drains for debris blockage or clogs. Check fittings and splices for tightness and make sure seals are intact. Is there any evidence of ponding sediment?

Damage: Look for and carefully document any signs of material distress caused by wind, thermal action, roof traffic, or equipment installation. Check eave, rake, and ridge lines, side seams, end laps, and penetration curbs, flashings, and trim.

Mechanical: Check all mechanical venting apparatus for proper operation and weather integrity.

Fasteners: Look for loose or missing fasteners. Note their location. Washer seals should exhibit uniform compression. There should be no splits, cracks, or spinouts.

Closures and sealants: Check closures and sealants at all joints for signs of moisture, sun, or contaminant degradation. Materials should be flexible and resilient. There should be no apparent voids, cracks, or insulation exposure.

Roof sheets: Examine the sheet surface condition for perforations, coating peels, cracks, sediment deposition, staining, and corrosion. Note the location of any aberration.

Stains and corrosion: Carefully document any evidence of staining or corrosion. Meticulously inspect exposed fasteners and sheet metal seam welds. Look for runoff or condensate staining. Check walls for gutter leak stains. Examine dissimilar material interfaces for sacrificial corrosion. Is there any corrosion around vent stacks? Are finishes at cut metal edges intact?

Insulation: Look for interior bagging or drooping and any evidence of moisture absorption. Keep accurate heating and cooling records. Energy efficiency losses may be traceable to insulation saturation.

Moisture damage: Look for signs of masonry efflorescence or cracks and floor, wall, and ceiling stains or peeling paint.

Once the inspection is completed, notify the installer of any conditions that would appear to be detrimental to long-term performance.

5.8   FAILURE DETERMINATION

The following list identifies areas about which information must be gathered to perform a comprehensive leak analysis. Subsequent evaluation of the information by a qualified professional may then provide clues as to the nature of the failure. This information will also aid in the assessment of leak potential as it relates to given sets of circumstances. Construction plans, specifications, and installation instructions are also extremely valuable assets to a professional analysis.

General Data

Building use	Leak history
Roof construction period	Leak maps
Roof design capacity	Thermographic scans
Roof sheet layout	Owner maintenance program

Geographic Data

Building orientation	Prevailing wind direction
Atmosphere (industrial, marine, etc.)	Violent weather history
Climate data (temperature, barometer, wind, rain, etc.)	

Structural Data

Primary framing layout	Static and dynamic loads
Secondary (roof) framing layout	Roof system type
Building bracing mechanisms	Roof sheet gage

Interior Conditions (Fig. 5-8)

Insulation type and thickness	Leakage evidence
Vaporretardertypeandperm rating	Condensation evidence
Insulation condition	Water damage (location and severity)
HVAC type and venting	
Settlement cracks	Underside sheet condition (rust)

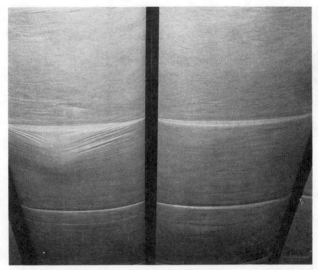

Figure 5-8. Sagging of blanket insulation due to water accumulation on top of the vapor barrier.

Roof Surface Conditions (Fig. 5-9)

Slope run and length
Drainage type and run
Drainage system
Drain damage
Ponding evidence
Seam and lap orientations
Straightness of architectural
   (eave, ridge, etc.) lines
Roof sheet modularity
Side seam integrity
Eave, valley, and rake line
   integrity

Through-roof penetrations
Roof sheet pillowing, draping,
   and depressions
Types, locations, and dimensions
   of penetrations
Side seam waver
Membrane (wind, hail, traffic,
   thermal, etc.) damage
Staining
Corrosion
Debris
Membrane damage or repairs

Sheeting (Fig. 5-10)

Eave, lap, valley, side seam,
   rake, ridge, and hip joint
   assembly (location, alignment,
   fit-up, edge overlap, fasteners,
   reinforcing plates, and
   sealants)

Rib tilt or stretch
Flat warping or dishing
Differential expansion/
   contraction restraint
Incompatible material contact
Damage or corrosion

Figure 5-9. Ponding sediment at every membrane end lap joint.

Figure 5-10. Flat dishing at eave line resulting in localized ponding at eave fasteners.

Figure 5-11. Flashing slopes toward building wall. Standing water at transition.

Flashing and Trim (Fig. 5-11)

Assembly (alignment, fit-up, splices, returns, fasteners, and sealants)

Material incompatibility

Transition seal integrity and flexibility

Drainage slope

Expansion/contraction restraint

Damage or corrosion

Penetrations (Fig. 5-12)

Upflow, downflow, and edge lap splice assembly (fit-up, edge transition, closures, fasteners, and seals)

Transition flashing and seals

Flow diverter restriction

Plugged drain channels or siphon breaks

Seam weld cracks

Sealant damming

Direct, through-accessory penetration

Profile damage

Differential expansion/ contraction restraint

Material incompatibility

Corrosion

Cracks

Mechanical unit condensation or leakage

Figure 5-12. Inadequate diverter installation causing upslope damming.

Figure 5-13. Improperly driven eave fastener. Corroded hole.

Fasteners (Fig. 5-13)

Proper type

Location and pattern

Engagement (stripped, jacked, or tilted)

Missing, broken, or overdriven

Shaft slotting or tilting

Corrosion

Rubber Closures (Fig. 5-14)

Type

Location

Fit-up

Compression (voids or gaps)

Mastic separation

Material degradation (exposure)

Elasticity

Material compatibility

Sealants (Fig. 5-15)

Type

Placement

Returns (pig tails)

Voids or gaps

Elasticity

Material compatibility

Material degradation (exposure)

Adhesion

Remedial applications

5.9   MAINTENANCE PROCEDURES

If designed and erected properly, a structural metal roof system will provide years of dependable service with very little physical maintenance. Many of the maintenance procedures are nothing more than a logical extension of the recommended periodic inspection process. If roof debris is present, clean it up; if drains are blocked, clear them; if drain seals are suspect, have them resealed; if accessory finishes are deteriorating, restore them; if fasteners are missing, loose, damaged, or corroded, replace them (replacement should be made with a fastener of compatible mechanical and finish properties that has a slightly thicker shaft than the original to ensure positive material engagement—i.e., a strip-out fastener); if mechanical accessories are defective, have them repaired or replaced; if HVAC runoff or condensation staining is found, provide a means of collection and channel it off the roof through plastic piping; if gaseous emission corrosion is found, raise the stack height and have the corroded surface restored.

Small voids in joint closures and sealants can be repaired with like materials. If sheet or flashing and trim joint seals display noticeable distress, notify your roofing consultant immediately (especially if wind or thermal action is the suspected cause). Ponding is another cause for substantial concern. There are no routine procedures or quick fixes that will resolve this condition. Document the occurrence as thoroughly as possible and call in an expert.

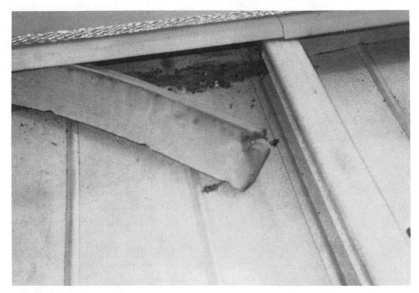

Figure 5-14. Poor closure fit-up resulting in transition joint seal deterioration.

Figure 5-15. Sealant dam at end lap. Splice joint.

Small, mechanically caused membrane perforations may be patched. Material compatibility must be maintained in order to ensure long-term effectiveness. Deterioration of the sheet finish is another problem. Specifying an improper restoration coating serves only to mask a sustained problem. At worse, it may accelerate it. If the problem is basically one of serviceability or if it is cosmetic in nature, two kinds of topical solutions are available: paints (film) and elastomeric sealants. The cause of membrane deterioration should be accurately assessed prior to the implementation of a topical application.

There are a number of items that must be evaluated when considering a film-type restoration. The most important is the determination of what impact the procedure has with respect to the original material and installation warranties covering the roof membrane. There may be specific provisions governing the touching-up or refinishing of the original surface. If a film application is determined to be an acceptable and viable restoration procedure, there are a number of other variables that it would be wise to examine before a final selection is made. These include the composition of the substrate material (zinc-, aluminum-, or alloy-coated), the necessary preparation of the substrate prior to application of the restorative film, and the composition and application methods of both the primer and the finish coat. In order to ensure compatibility with an existing coating, the same information must be known about the original factory finish.

The proposed film coating should also be evaluated regarding adhesion; flexibility; chalk, fade, and mar resistance; permeability; hardness; and moisture, chemical, and corrosion protection. Comparative values can be obtained from manufacturer laboratory test reports.

If an elastomeric sealant coating is being considered, the preliminary evaluations are similar to those of the film. In this case, the composition and proper preparation of the existing surface is paramount to the success of the restoration. If the original surface is not correctly prepared, anything put over it will fail regardless of the quality of the applied product. Once the remedial coating fails, the owner is effectively left with one of two choices: reroof or retroroof.

Sealant coatings are typically applied in 20–60 mil thicknesses. They are supposed to enhance moisture protection as well as extend membrane surface durability. Things to look for in a sealant coating product include substrate adhesion, elasticity (remember that the metal panel will continue to expand and contract at a different rate than the coating and the coating must be capable of absorbing this relative movement without delaminating), atmospheric and chemical resistance (including ultraviolet and ozone exposure), physical toughness, and fire resistance. Never allow an asphaltic or asphalt-based coating to be applied to a metal roof. Even spot treatments will fail in a short period of time.

Both film and sealant coatings are highly proprietary. Installed costs may vary substantially with product type and required substrate preconditioning. Life cycle costs of topical coating applications should be compared with those of more permanent alternatives.

5.10   LEAK REPAIR

No roof system is infallible. Even a quality installation can develop an occasional leak during the course of its service life. It is extremely important that any leakage, no matter how minor, be investigated immediately and repaired as soon as possible. A leak signals a breach in the system. It will allow moisture penetration into concealed cavities and, ultimately, degradation of building materials. Aside from a certain loss in thermal performance, and the possible development of progressively worsening condensation problems, entrapped moisture may react chemically with the insulation and other building materials. The reaction may actually result in the formation of weak acid solutions that could attack and undermine the material integrity of the structural support system.

Many owners attempt a multiplicity of repairs in an effort to cheaply alleviate minor leak problems. Only in the face of repeated failures does the owner acquiesce to bringing in a professional. The expert is then left to decipher a maze of conditions that may not remotely resemble the original installation. The real problem may lie masked under layers of ineffective gun grade caulk and tar and may even have been compounded by them. It is not unusual for a situation that has been allowed to progress to this point to be irreparable. The owner must then make the hard choice of either living with the situation or commissioning a complete restoration or retrofit (Fig. 5-16).

By obtaining the following data immediately upon discovery of a leak, the owner's chances of quickly and effectively solving the problem are dramatically increased:

- Pinpoint the exact interior location of the leak as close to the underside of the roof surface as possible. Once moisture penetrates the exterior membrane surface, it will "run" through and along other building materials and equipment until an exit or drop point is reached. The point where the drop hits the floor or a ceiling tile may be far removed from the actual breach.
- Note the weather conditions bracketing the initiation of leakage. Include precipitation amounts, wind velocities and directions, barometric pressures (both outside and inside the building), and interior and exterior temperatures and humidity.
- Describe the leak activity (started simultaneously with rain, started X minutes after rain started, continued X minutes after rain stopped, etc.).
- Check for and note the exact location of any localized ponding on the roof

Figure 5-16. "It's leaking somewhere around here."

surface. Does the ponding occur only in apparent low spots or does there appear to be damming due to wind or surface obstructions? Is there ponding around roof drains? Is there evidence of gutter overflow? When the rain stops, do the ponds dry up or drain out?
- Check for any obvious wind movement or distortion of sheeting, flashing, and trim material, especially at building corners and roof surface elevation changes.
- Notify a qualified metal roofing contractor immediately and get notes and records ready.
- Do not attempt to make any roofing repairs independently except for temporary repairs in emergency situations.

More elaborate methods may be used to determine the source of a leak, such as thermographic mapping, joint disassembly and inspection, insulation acidity determination, and weeping suspect locations. The owner should discuss the value of these extra measures with a roofing expert.

5.11 RETROFIT ROOFS

Retroroofing is an extension of the metal roof concept into the maintenance market. It has proven itself to be a workable and highly attractive alternative to refurbishing both existing conventional and metal roofs that have reached the end of their serviceable life. A retroroof is essentially an over-roof of the standing seam variety. It generally does not require removal of the existing roof, as a replacement roof would, providing that the existing membrane and its underlying support elements are structurally sound. Gravel from built-up roofs and ballast from loose-laid roofs are typically removed to reduce dead weight and to simplify the erection of the new system. All previous considerations applicable to the design and selection of a new metal roof apply to retrofits as well. Since this installation must mate with an existing set of conditions, there are several additional factors to be accounted for:

- Site logistics associated with the proposed reconstruction. It is likely that the installation will have to be accomplished without interruption to the facility's normal business activities. Planning is critical. Items to be resolved in advance include access for construction equipment, access for material unloading, staging and preassembly areas, access for utility and service hookups, provisions for temporary moisture protection and drainage, provisions for preservation of exterior building elements not purposefully affected by the new construction (wall surfaces, building accessories, and equipment), protection of critical interior processes (computer facilities, precision manufacturing), and interface capability of existing structural conditions.
- Capability of assimilating existing roof surface conditions into the retroroof system. The conditions include profile variance, site drainage plans, roof-mounted accessories (skylights, HVAC equipment, and process lines), and surface perimeter architectural conditions (parapets and facades). Other less obvious, but equally important, considerations include building and fire code restrictions, safety, noise, and the affects of the alteration of snow and wind load patterns on the existing support structure.
- Moisture removal from the existing roofing materials. This procedure will ensure against long-term thermal efficiency breakdowns or condensation problems. If a new attic space is created by the retroroof installation, it may have to be vented for similar reasons. The proposed retrofit roof system must accommodate all of these considerations.

A new metal roof installation may run $2–$4/ft² depending on type, material, gage, finish, roof profile, insulation system, and locality. A retrofit roof may run from $3 to $20 per square foot (installed) depending on the complexities of and obstacles to construction. It pays to do some careful shopping when considering a retroroof. As with all metal roof systems, look for a complete, single-source package.

5.12   SUMMARY

There is no single best roofing choice for all situations. Advanced metal roofing technology has established itself as a formidable force in the roofing industry. It is a viable and cost-effective solution for many situations. Of the two classes of metal roofing available—architectural (a topical decorative application) and structural (a multifunctional performance-oriented element) —strength, durability, and versatility have vaulted the structural roof to the forefront of the industry.

The most recent development in structural metal roofing technology is the SSR. Its design has many advantages over its long-term predecessor, the through-fastened or screwed-down metal roof. The SSR can be used on slopes as low as ¼ in./ft, is adaptable to more efficient insulation systems, and is virtually immune to thermal shock distress. It is a floating roof system, with no diaphragmatic capability for resisting lateral loads or forces (as a screwed-down assembly has). SSR's independence of the substructure yields performance advantages, but also makes it more complex and slightly more expensive than through-fastened roofs.

Good design will eliminate much of the failure potential associated with metal roof systems. Simplicity and attention to synergistic details are keys to successful installation. Insistence on certain desirable proprietary design features such as factory-fabricated laps, machined interlocks, and UL90-rated assemblies will also enhance the chances for success. Then, by adding a reputable product and an experienced qualified installer, a problem-free roof is virtually assured.

Roof system failures provide over 70% of the litigated claims brought against the construction industry. Performance failures are usually traceable to manufacturing, design, and installation errors, as well as owner abuse. Most failure cases are a result of a combination of these errors.

Nearly all leaks are the result of joint failure. Because joint assembly techniques vary with location (eaves, rakes, ridges, hips, elevation breaks, end laps, and penetrations), it is often difficult to isolate a potential cause of failure. Organizing an investigation with the help of a guideline checklist will lead to a more expedient solution than a caulking gun used to shoot every joint full of sealant.

The owner can extend the performance integrity of a quality installation many years by instituting a program of regular inspections and light maintenance. Inspections can be professionally contracted or delegated to the facility's maintenance crew. Professional services may include everything from examination to all-inclusive maintenance and repair. Whether contracted for or performed internally, the inspection should include checks for roof debris, inhibited drainage, material damage, rooftop accessory and equipment malfunction, joint seal integrity, staining, corrosion, insulation damage, and moisture intrusion. Internally delegated roof maintenance should be limited to inspection and general cleanup. A qualified installer should make all needed repairs.

Even a quality installation may develop a leak during its service life. A leak is a breach in the system and must be addressed immediately. Accurate and specific documentation of the incidence and the atmospheric conditions at the time of the leak will significantly aid in the determination of the leak source and its repair. If required, more sophisticated detection techniques such as thermographic scans are also available through a qualified roofing expert.

A properly designed and erected structural metal roof system will provide years of dependable service and demand little in the way of actual physical maintenance. The impenetrable roof has not yet been developed. No one solution has demonstrated dominant performance under all circumstances. Structural metal roofs, however, offer a uniquely versatile and cost-effective answer to a myriad of concerns related to thermal and moisture protection.

5.13   SUGGESTED READINGS

Fricklas, Dick. 1986. Structural vs. architectural metal roofs. *Exteriors* 4(2, summer):8.
Low Rise Building Systems Manual. 1986. Cleveland: Metal Building Manufacturers' Association.
Nimtz, Paul D. 1987. Comparing structural and architectural metal roofing. *Exteriors* 5(4, winter):46–49.
Nimtz, Paul D. 1987. Regular inspections important to maintaining standing seam roofs. *Metal Architecture*, November, p. 8.
Reference Guide to Metal Roof Systems, September 1988 issue of *Metal Architecture*.
Stephenson, Fred. 1986. Metal systems offer both design, performance benefits. In *Handbook of Commercial Roofing Systems*. Duluth, MN: Harcourt Bruce Jovanovich.

5.14   ASSOCIATIONS WITH INFORMATION ON EXPOSED METAL ROOF SYSTEMS

American Iron and Steel Institute
1133 15th Street, NW Suite 300
Washington, DC 20005-2701

Institute of Roofing and Waterproofing Consultants
51 West Seegers Road
Arlington Heights, IL 60005

Metal Building Manufacturers Association
1230 Keith Building
Cleveland, OH 44115

Metal Construction Association
1133 15th Street, NW Suite 1000
Washington, DC 20005-2701

Roofing Communications Network
1720 West End Avenue, Suite 601
Nashville, TN 37203

Roofing Industry Educational Institute
7006 South Hilton Way #B
Englewood, CO 80112-2003

6. Concrete Surfaces

James Warner

6.1 INTRODUCTION

Concrete is one of the most versatile materials available to the builder. Composed of portland cement, sand, large aggregate and, in some cases, chemical admixtures to enhance its workability or other properties, it can be cast into virtually any shape or configuration. Whether cast in a horizontal configuration and finished by troweling, or placed in temporary forms, as in the case of vertical or overhead castings, it can be given a nearly infinite number of surface textures or finishes.

With quality design, proper construction, and reasonable maintenance, structures built of concrete can last nearly forever. In fact, early concrete structures including sections of the Roman aqueduct exist to this day. Likewise, in the United States, many very old historic concrete structures remain serviceable. Some examples include the Milton House in Milton, Wisconsin (1844), the Sebastopol House in Seguin, Texas (1856), and the Ponce de Leon Hotel in St. Augustine, Florida (1909).

Notwithstanding the longevity and good performance of so many concrete structures, similar structures lacking proper design, construction, and/or maintenance have deteriorated rapidly. The records are filled with such examples, often citing deterioration so extensive that demolition of the structure was required. Accordingly, proper maintenance is essential in order to obtain the enduring service that this unique material is capable of providing (Table 6-1).

The primary objective of this chapter is to deal with the maintenance aspects of concrete structures. However, an understanding of the material, including design and construction aspects that affect the structure's eventual durability, is also important.

Table 6-1.   Interacting Factors That May Induce Premature Deterioration of Concrete

Premature Deterioration Factor	Characteristic of the Concrete	Characteristic of the Environment
1. Freezing and thawing	Lack of entrained air in the cement paste or excessively porous aggregate, or both, in concrete critically saturated with water	Moisture, and freezing and thawing
2. Aggressive chemical attack		
a. Sulfate attack	Excessive amounts of hydrated calcium aluminates in the cement paste	Moisture containing dissolved sulfates in concentration
b. Leaching	Excessive porosity	Moisture of low pH and low dissolved lime content
c. Surface deterioration	Surface disintegration	Aggressive chemical exposure, usually with water
3. Abrasion	Lack of resistance to abrasion	Abrasive, often in or under water
4. Corrosion of embedded metal	Corrodible metal and (usually) corrosion-inducing agents in the concrete	Moisture (or moisture and corrosion-inducing agents)
5. Alkali-silica reaction	Excessive amounts of soluble silica in the aggregate and (usually) alkalies in the cement	Moisture (or moisture and alkalies)
6. Other		
a. Unsound cement	Excessive amounts of unhydrated CaO or MgO	Moisture
b. Plastic shrinkage	Lack of maintained moisture content during the time that the concrete remains plastic	High evaporation rate

Note: A periodic washdown of concrete surfaces with potable water can retard the rate of concrete deterioration if contaminants are from exterior environment.

*Prevent wetting if possible. Provide for rapid drainage. Do not allow water to pond.

Manifestation of Deterioration	Recommended Maintenance/Repair
Internal expansion, cracking and popouts	Take out and replace with frost-resistant concrete*
Internal expansion and cracking	Remove concrete and replace with sulfate-resisting concrete*
Dissolution and removal of soluble constituents	Remove concrete and replace with low-permeability concrete
Surface dissolution	Remove concrete to sound material and replace—consider chemically resistant replacement material
Removal of material	Remove concrete down to sound material and replace with abrasion-resistant concrete or other material
Internal expansion and cracking	Remove affected concrete exposing corroded rebars, remove corrosion product or replace rebars as necessary and repair with low-permeability concrete*
Internal expansion and cracking	Serious problems require removal of unsound concrete and replacement with concrete not susceptible to such reaction*
Internal expansion and cracking Cracking at very early ages	a and b: If cracking is not showing all through the concrete, a "breathable" sealer can be applied; otherwise, the unsound materials must be removed and replaced

6.2   COMPOSITION OF CONCRETE

Concrete is made by combining cement, water, and aggregate into a uniformly mixed substance. The combined cement and water are referred to as cement paste or gel. It is this paste or gel that binds the aggregate together. The binding action results from a chemical reaction between the cement and water known as hydration. Enough water must be provided to permit a sufficient amount of the cement to hydrate. Water in excess of what is required is referred to as surplus water and will result in shrinkage and excess porosity of the hardened mass.

Drying shrinkage of concrete continues for a year or more after initial casting and often results in excessive cracking. Concrete that has been properly proportioned and consolidated during placement will be impermeable to liquids. However, the permeability of concrete is directly related to its porosity. Additionally, porosity will adversely affect other important properties of the composition such as strength, durability, resistance to abrasion, and resistance to freezing and thawing damage.

It is virtually impossible to provide too little water in most conventional concrete mixtures. The amount of water required to obtain the necessary plasticity and workability for proper placement is generally much greater than that required for proper hydration. Due to mixing and placing constraints, the concrete in virtually all structures contains more water than is desirable, resulting in a finished mass of greater porosity and permeability, and thus lower durability than would be ideal.

It follows that use of a concrete mixture that contains the minimum feasible amount of water should be considered. Many well-established methods for such mix designs are described in *Design and Control of Concrete Mixtures* (Portland Cement Association, 1977). Concrete mixtures designed for minimal shrinkage and porosity should have a low total water content and should contain the largest practicable amount of aggregate consistant with the required workability. Also, the largest-size aggregates that can be handled and placed should be used, as mixes containing large-size aggregates have more total aggregate, and thus they have less pore space requiring cement paste than those containing small-size aggregates.

Shrinkage, which usually continues for more than a year after casting, is also influenced by the type of aggregate. Hard aggregates are difficult to compress and therefore provide more restraint regardless of shrinkage of the cement paste. A variety of water-reducing admixtures are available commercially and such materials have been in use for many years. These admixtures are used to reduce the quantity of mixing water required to produce concrete of a given consistency, without changing the amounts of any of the other constituents.

In addition to water-reducing admixtures, additives are available to obtain other desired properties. Set-accelerating and set-retarding admixtures allow control of the time that the plastic concrete will remain workable. Also available are air-entraining admixtures that will provide a matrix of closely spaced minute air bubbles in the concrete as required for resistance to freezing and thawing.

Hydration and thus formation of the cement gel starts as soon as the cement comes in contact with water. It continues as long as water is available or until 100% hydration has occurred. In actual practice, 100% hydration virtually never occurs. Likewise, availability of water to continue the hydration process, commonly referred to as curing, is seldom, if ever, long enough. So maintenance requirements of concrete structures are affected greatly by the quality of the concrete mixture and the length of time and quality of curing during the original construction.

6.3   TYPES OF DEFICIENCIES OF CONCRETE

Concrete deficiencies can be divided into two distinct groups: structural failure and lack of durability. Structural failures are usually the result of a design deficiency or an overload condition, either natural, such as an earthquake, or manmade. When this type of problem is experienced, the required remedial effort is far beyond maintenance. Structural failure requires appropriate analysis and redesign. Under such conditions, a fully qualified structural engineer should be promptly retained.

There are many possible causes for lack of durability of concrete. It is interesting to note, however, that a major contributor to practically all durability problems involves permeation of the concrete with water, other liquids, air, or gas. Typical problems are discussed below.

6.3.1   Freeze-Thaw Damage

Scalling of the surface concrete (Fig. 6-1) occurs when water is able to fill capillary voids within the concrete mass prior to freezing. When water freezes it expands with such force that failure and disruption of the overlying concrete occurs. To be susceptible to freeze-thaw damage, the concrete must be sufficiently permeable to water in the first place. Freeze-thaw damage can be prevented by producing sufficiently impermeable concrete so saturation with water is not possible, providing proper air entrainment with a suitable admixture, or both. Low permeability of the concrete can be assured by utilizing aggregate that is low in pore volume, permeability, or both, and minimizing the quantity of mix water used.

Figure 6-1. Surface scaling resulting from freezing when concrete is saturated. Such damage is aggregated by ponding of water.

As an alternate to providing an impermeable cement paste (possible but infrequently obtained with the current state of practice), an appropriate air-entraining admixture should be used to provide a paste with a proper spacing of minute air bubbles. Inclusion of such an air void system has been proven to provide frost-resistant cement paste and its use is well documented in *Concrete Manual* (Bureau of Reclamation, United States Department of the Interior, 1981). If the freezing water can reach an air void or bubble before the rising pressure becomes excessive, damage will be prevented. Air-entraining admixtures are formulated to provide such relief.

6.3.2 Chemical Attack

Aggressive chemical exposure can result in either disruption of the concrete mass due to internal expansion or disintegration of the mass by dissolution and loss of chemically soluble constituents. A common type of chemical attack is exposure of concrete to soil or water containing excessive concentrations of sulfates. Obviously concrete that is maintained in an aggressive environment, especially low pH (acidic) as in process chemical plants and some food-handling facilities, is prone to acid attack.

6.3.3 Alkali-Silica Reaction

Aggregates with particular mineral constituents such as chalcedonic cherts and some dolomitic limestones become reactive when in contact with excessive amounts of alkali. The reaction manifests itself through internal expansion of the concrete mass and resulting cracking or surface popouts. In order for the alkali-silica reaction to occur, three elements must be present: (1) a potentially reactive aggregate, (2) a sufficient source of alkali, and (3) water. Many aggregates, although potentially reactive, have been used successfully with relatively low-alkali cement. The incidence of alkali-silica reaction increased in the United States as a result of the cement shortages of the 1970s and, in more recent times, due to the use of nontraditional (imported) sources of relatively high-alkali cement. It is thus important to confirm that both the cement and the aggregate proposed for use are compatible with each other.

6.3.4 Corrosion of Embedded Metal

Perhaps the most frequent problem encountered with concrete structures is cracking, spalling, and delamination of the surface as a result of corrosion of the embedded reinforcing steel or other ferrous metals. Concrete normally provides an ideal environment for metals as its high alkalinity provides a passive oxide film around the embedded metals protecting them from corrosion. Corrosion occurs when this passive film is broken. Such interference can be initiated by a general lowering of the pH of the surrounding concrete (a pH of around 11 is required for passivity), or by the presence of corrosion-promoting chemicals such as chlorides.

The corrosion process depends on an electrical current between the anode (the area where the actual corrosion occurs) and the cathode (another location on the same piece of steel or any other metal that is in electrical continuity with the anode). To complete the electrical circuit, the current must flow through the concrete. All concrete, unless it is oven-dry, is conductive to some degree. The degree of conductivity, however, is directly related to the moisture content and thus permeability of the concrete. When corrosion does occur, the corrosion that is produced occupies a greater volume than the material from which it is derived. Expansive forces, which are greater than the tensile strength of most concrete, are developed during that process, thus fracturing the concrete.

6.3.5 Carbonation

Another cause of corrosion of embedded metal is carbonation, resulting from exposure to excessive concentrations of carbon dioxide in the atmosphere. This later cause is not much of a problem in most areas of the United States where carbonation rarely penetrates more than about $\frac{1}{4}$ in. from the surface of the concrete. However, in parts of Europe, particularly the United Kingdom, carbonation of more than 1 in. in depth has frequently been reported.

The best protective measures against corrosion of embedded metals are to provide adequate concrete cover over reinforcing steel and to ensure that the cover concrete is of the lowest permeability practicable. For new construction, chemical admixtures that inhibit the action of chlorides are available as is fusion-bonded, epoxy-coated reinforcing steel, which insulates the steel from the concrete and thus prevents corrosion. Use of these special materials should be considered when casting new concrete in aggressive environments.

6.3.6   Mechanical Abrasion

Most normal concrete surfaces are very resistant to abrasion resulting from ordinary exposures. However, under extreme exposure to abrasive forces, such as small steel-wheeled carriages, rocks, or other debris transported in rapidly moving water, surface erosion can occur. The susceptibility to such abrasive wear is directly related to the quality and hardness of the surface layer of concrete. Therefore, particular attention to the concrete quality is essential in the construction of any work that will be exposed to unusual abrasive elements. In an extraordinary case, the surface can be protected by a surface hardener, generally dusted onto the surface and troweled in during construction, or the application of an abrasive-resistant polymer coating or overlay.

6.3.7   Nonuniform Volume Changes

As with all materials, concrete is subject to volume changes with fluctuations of temperature. These properties are well understood and are usually provided for in design and construction through realistically sized sections and adequate joints for movement. Where adequate jointing has not been provided, the concrete will form its own joints by random cracking. It can also cause distress in existing joints or bearings (as in the case of bridges and precast structures). Possible solutions for nonuniform volume change problems include control of the temperature by shading with canopies, awnings, or other enclosures, environmental temperature control, or provision of adequate movement joints.

6.4   MAINTENANCE CONSIDERATIONS DURING DESIGN AND CONSTRUCTION

Virtually all problems associated with concrete surfaces are influenced by the porosity and permeability of the surface concrete. Further, most problems involve the penetration of water into the concrete. For this reason, serious consideration must be directed toward minimizing the ability of the concrete to absorb moisture. Sufficient slope for positive drainage of concrete surfaces should always be provided. Water should not be allowed to drain onto horizontal concrete surfaces such as roof drains placed over lower projections or bridge deck drains placed over piers. Horizontal surfaces should slope a very minimum of ⅛ in./ft and preferably ¼ in./ft of run. Parapets, cornices, sills, and all projections from the structure should likewise be designed for positive drainage. The importance of careful detailing to provide positive drainage of all concrete surfaces cannot be overstressed.

Also of great importance is providing a sufficient amount of low-permeability cover concrete over all reinforcing steel or other embedded metals. Low-permeability concrete will absorb moisture less rapidly than higher-permeability concrete and thus offers significantly greater protection. Close control of the concrete mix is in order, with particular attention paid to maintaining a low water-to-cement ratio and low total water content. Where aggressive environments are to be encountered, adequate cover over embedded metals should be ensured. The amount of such cover required will vary depending on the particular exposure that is likely to occur. A guide to reinforcing steel coverage requirements will be found in the *Building Code Requirements for Reinforced Concrete* (ACI 318-83, American Concrete Institute, 1983).

In extreme cases, application of a surface-protective coating may be in order. Select the coating with care as the concrete must be able to breathe. Generally speaking, protective sealers and coatings should be impermeable to water or other liquids, but permeable to air or gas. In extreme environments, it might be advisable to specify epoxy-coated reinforcing steel and corrosion-inhibiting admixtures. Obviously, in such extreme conditions, ample cover concrete and concrete of high quality (low permeability) must be used.

Rational spacing and placement of reinforcing steel and other embedded items such that there is minimal interference with concrete placement is essential. Also, members should be designed so that their configuration does not interfere unreasonably with concrete placement. Where reasonable access for placement cannot be provided, special methods of concrete composition should be considered (Warner, 1984).

The provision for thorough and effective curing of concrete should be mandated. Due to the near impossibility of application of uniform coverage where liquid curing membranes are allowed, heavier coverage than that recommended by the manufacturer should be considered. Likewise, where severe environments are encountered, the requirement for extended curing periods as well as more effective curing methods such as continuous ponding should be considered.

As with all building materials, concrete expands and contracts with changes in temperature. Additionally, it shrinks as it cures. All concrete structures must therefore be designed to accommodate such changes. Adequate movement joints must be provided. Failure to make such provision will result in the formation of random cracks.

In addition to following through on the above items during construction, special attention should be directed to formwork and concrete placement. Formwork for concrete must not only be constructed to true and accurate dimension but it must be sufficiently tight so that excessive leakage of water does not occur. As previously discussed, virtually all concrete contains surplus water when it is placed. If the formwork is tight, most of this surplus water will go to the top of the mass prior to its hardening. Such water is referred to as bleed water, and it results from settlement of the more dense constituents of the mix. However, if open joints or holes exist in the formwork, much potential bleed water will tend to leak from them. When such leakage occurs, cement will tend to combine with the leaking water, which will result in sand streaks of other surface blemishes deficient of cement.

Concrete placing methods used should provide for placement without segregation of the mix constituents. In reinforced sections, the concrete must not be allowed to free-fall so the large aggregate is separated from the mix as it falls on the reinforcing steel. This requirement often dictates that the concrete not be allowed to free-fall a vertical distance more than 3–4 ft in heavily reinforced sections. As the concrete is placed, it must be consolidated such that it thoroughly and completely fills all spaces around the reinforcing steel and any other embedments, is free of any rock pockets, and is compacted into a dense mass free of voids or air pockets. Rodding, spading, or tamping with hand-held implements can be utilized, although vibration is the most effective and most frequently used method.

Vibration can be accomplished with either hand-held "spud" vibrators that are immersed into the concrete as it is placed or external vibrators that are attached to the formwork. The most common form of consolidation is with internal or immersion-type vibrators. There are several types of vibrators available with varying operational characteristics. *Design and Control of Concrete Mixtures* (Portland Cement Association, 1977) provides a much more complete coverage of this important construction requirement.

The forms should remain in place until the concrete has reached a strength sufficient to be self-supporting and until a surface hardness has been reached that will withstand form removal without surface injury. For overhead elements, the concrete must reach a sufficient strength to be self-supporting prior to removal of the temporary supporting shores. Formwork that remains in place for a considerable time should be wet down periodically to aid in proper curing. Once the formwork is removed, it is imperative that proper curing procedures are immediately initiated.

6.5   IDENTIFYING POTENTIAL PROBLEM AREAS

Existing structures should be periodically inspected for early detection of potential problems. It is a good idea to make inspections immediately after a rain storm so problems of ponding and poor drainage can be easily detected. Of special concern should be ponding of water, a condition that should be remedied immediately. This solution can be readily accomplished by chipping, grinding, or otherwise cutting a drainage groove in the concrete to provide positive drainage, or by filling in the low area. It is sometimes judicious to provide pipe or other drains or to reroute the drainage flow patterns. Sources of drainage water should be observed, and where possible, the water should be rerouted so as not to drain onto or cause ponding on concrete surfaces. Often such potentially damaging conditions can be easily remedied with relatively minor effort.

Visually examine surfaces for distress such as rust stains, unusual cracking, spalls, or other surface defects. If significant, these need prompt attention. Such defects will often signal the early stages of much greater problems, which likely will require action far beyond normal maintenance. For this reason, it is usually advisable to obtain the services of a qualified engineer to evaluate the problem and make recommendations for any remedial action. Early detection and corrective action for such defects cannot be overstressed.

6.6   REPAIR MATERIAL SELECTION

Selection of appropriate materials is an absolute requisite for obtaining durable repairs. In order to match the properties of the base concrete, there can be no question that the best material for concrete repair is concrete itself, or other similar cementitious compositions. Unfortunately, such materials cannot always be used. Such factors as the requirement for rapid strength, restricted working conditions, improvement of chemical or abrasion resistance, and need to achieve aesthetically pleasing surfaces often require synthetic materials.

Each repair job has its own conditions and special requirements. It is only after these are thoroughly examined that the final repair criteria can be established. Once this assessment has been made, it will often be found that more than one material can be used with equally good results. Final selection of the actual material or combination of materials to be used then becomes a decision for the repair person to make. This decision must take into account cost, ease of use, and available laborer skills and equipment, as well as the properties of the proposed materials.

The maintenance engineer has an almost endless array of specialty, as well as conventional materials, with which to perform repairs. The easy availability of different materials provides many advantages, but also results in much confusion. Frequently inappropriate materials are selected, which leads to a large number of repair failures. While both bond and compressive strength values are often provided by material suppliers, more fundamental characteristics such as the material's dimensional stability, stiffness, and ability to transmit fluids, vapors, and electrical current are of equal or greater importance. The following discussion of these factors is provided to improve understanding of their importance and to aid in making the best material selection.

6.6.1   Dimensional Stability

Many concrete repair jobs require the formation of patches or other additions to the original concrete section. Virtually all concrete used in orthodox construction will shrink with time. Although the largest component of the total shrinkage will occur soon after the concrete is cast, significant drying shrinkage will usually continue for a year or more. Thus, cracks often occur a considerable time after casting of the concrete. The shrinkage potential of concrete (in new construction) is usually taken into account in design and construction by providing expansion and control joints. Because most repairs are not made until the concrete is of some age, and thus not subject to significant further shrinkage, the repair material must be essentially shrinkage-free, or potential shrinkage must be provided for in a manner that will not cause failure.

One of the primary considerations of any repair is the provision for good bond of the new material to the substrate. Although there is a popular belief that new concrete cannot achieve a good bond to old concrete, it is not entirely correct. The lack of bond of new to old concrete is not caused by the inability of the new concrete to adhere, but rather by shrinkage of the new concrete away from the old concrete that has already undergone most of its potential shrinkage. The excess shrinkage of repair preparations can be mitigated by using relatively shrinkage-free materials or by following construction procedures — such as drypack, shotcrete, preplaced aggregate concrete, or vacuum water removal of the repair section — that result in little or no shrinkage potential.

To combat shrinkage problems, many repair products are available that do not shrink or that actually expand in volume once they are mixed. Exercise caution when using expansive materials, as many must be fully restrained once they are placed to prevent excessive swell, resulting in loss of both strength and durability. Also, when considering use of an expansive material, determine the time period of expansion to assure that the repair can be completed before it occurs. When evaluating expansive repair materials, both restrained and unrestrained test specimens (Fig. 6-2) should be made. Strength is determined from the restrained specimens, whereas expansion potential is determined from the unrestrained specimens.

6.6.2   Modulus of Elasticity

The modulus of elasticity, sometimes called Young's modulus or simply "the modulus," refers to the stiffness of a material. A material of low modulus will tend to deform when excessive load is applied, whereas one of high modulus

Figure 6-2. Expansive grout specimens. Specimen on right has been restrained to be used for strength evaluation; specimen on left is left unrestrained for evaluation of the amount of expansion.

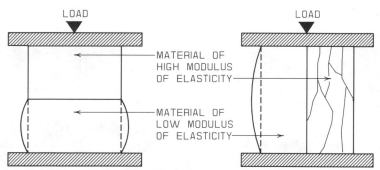

Figure 6-3. Modulus of elasticity. Low-modulus materials yield more than higher-modulus materials when loaded. If composite is vertically oriented all load is taken by higher-modulus portion, which can result in failure thereof.

will react with brittle fracture. While this factor may be of little consequence where materials of widely differing moduli are used in horizontal layers, when used in a vertical configuration, serious distress can result upon loading. As illustrated in Figure 6-3, the lower-modulus material will tend to yield or bulge, directing all of the load into the high-modulus material, which when overloaded, will fracture or shatter. This behavior is particularly evident where dynamic forces exist, such as under or adjacent to large machinery and in foundations and floors subject to impact or vibration. It also occurs when structures are subjected to natural dynamic forces, such as wind or earthquake shaking.

Very stiff patches should not be made within the mass of a relatively soft wall or other construction. Where it has been done, as in the case of using concrete to fill in a window opening within an unreinforced brick wall, the wall can literally be destroyed if subjected to excessive vibration or shaking, when the relatively soft brick is forced to yield around the stiff concrete. Likewise, when materials of widely differing moduli are placed within the same horizontal layer, such as steel shims within the grout under a baseplate, the lower-modulus material will likely yield and all forces are transmitted through the stiffer material. Figure 6-4 illustrates this point; the lower-modulus grout has nearly completely disintegrated, placing all of the load onto the very stiff shims that were improperly left in place. As can be seen, the resulting large unit force has caused failure of the underlying concrete substrate.

Where individual structural elements such as columns or beams are enlarged or repaired with materials of stiffness greatly differing from that of the original element, serious distress or complete failure can result. Figure 6-5 illustrates such a case. The existing concrete column was enlarged in order to increase its load-carrying capacity. In this instance, the additional material was an epoxy mortar with a much lower modulus of elasticity than the concrete

Figure 6-4. Spall failure of concrete caused by concentrated load transferred into it through much higher-modulus steel shim. Grout that was of much lower modulus than either concrete or steel has largely disintegrated, leaving partial void under base plate.

Figure 6-5. Concrete column that was strengthened by addition of relatively low-modulus epoxy mortar. Under load, higher-modulus concrete has shattered as lower-modulus epoxy mortar simply yielded rather than provided support for the load.

of the original column. When a full-scale specimen was removed from the structure, taken to the testing laboratory and load tested, upon loading the epoxy mortar merely yielded, forcing all of the load into the original concrete, which then failed completely.

6.6.3   Coefficient of Thermal Expansion

All materials expand and contract with changes of temperature. For some materials, the change in volume is very small, whereas other materials exhibit large volume changes. The rate at which thermal volume change takes place is the coefficient of thermal expansion. It is usually measured in millionths of an inch of volume change per linear inch of the material per degree of temperature change. When making large or thick patches, or additions to a concrete section, it is important to use a material with a coefficient of thermal expansion similar to that of the base concrete. When a composite of two materials of widely varying thermal coefficients undergoes a significant temperature change, a failure either at the bond line or within the section of the lower-strength material will occur, due to the differences in volume change (Fig. 6-6). Obviously, this factor is of greater importance in environments that experience significant temperature differentials.

However, not just the normal environment but all environments to which a particular area will be subjected must be carefully considered. For example, the temperature of freezers is periodically raised for routine cleaning and maintenance. Such changes of environment have resulted in many failures of otherwise good-quality repairs, especially of surface overlays in freezers and in other facilities that might be subject to large differences in environmental temperature. Figure 6-7 shows an exterior column of a building in the northeastern part of the United States, an area that experiences large variations in temperature. Shown is a patch of epoxy mortar that failed in less than one year. The failure is the result of combining materials of two widely differing

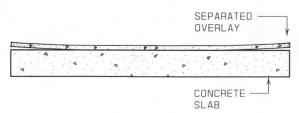

SEPARATED OVERLAY

CONCRETE SLAB

Figure 6-6. Typical failure due to incompatible coefficients of thermal expansion.

thermal coefficients in an exposure that will be subject to large temperature differentials. A different type of failure is indicated in Figure 6-8. In this case an overlay of greatly differing thermal properties was placed on an exterior concrete surface. As a result of temperature variation, the overlay completely separated from the parent concrete shortly after being cast.

6.6.4 Permeability

Permeability refers to the ability of a material to transmit liquids or vapors. Good-quality concrete is relatively impermeable to liquids, but transmits vapors. If impermeable materials are used for large patches, overlays, or coatings, moisture or vapors that would otherwise penetrate through the base concrete could be entrapped. Entrapped substances can build up enough pressure to cause a failure either at the bond line or within the weaker of the

Figure 6-7. Failure of epoxy mortar spall repair due to different coefficients of thermal expansion of the epoxy mortar and concrete.

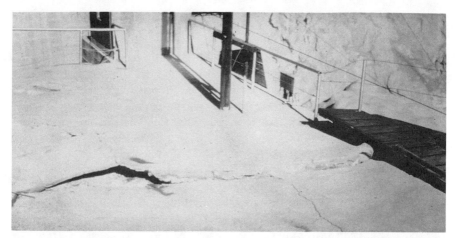

Figure 6-8. Failure of overlay due to different coefficients of thermal expansion of the base concrete and high-technology overlay material.

Figure 6-9. Failure of epoxy overlay caused by moisture vapor attempting to escape out of the slab-on-grade, but being unable to penetrate the impermeable epoxy.

two materials. This problem is common where impermeable overlays or coatings are placed on slabs on grade that are subject to freezing and thawing, placed on soil subject to saturation, or constructed on vapor barriers. Figure 6-9 shows the failure of an otherwise good-quality but impermeable epoxy overlay that has failed primarily due to the pressure of moisture vapor rising through the slab on grade.

Another area in which impermeable materials should generally be avoided is the patching of concrete that has been damaged due to corrosion of embedded reinforcing steel. As previously discussed, corrosion requires the flow of electrical currents through the concrete. Should a portion of the conductive base concrete be replaced with a nonconductive repair material, the transmission potential of electrical current will be changed. Such change often results in changes of the corrosion pattern, often accelerating corrosion.

Figure 6-10 shows a beam that had been repaired with nonconductive epoxy mortar less than a year prior to the time the photo was taken. After knocking off the repair mortar (Fig. 6-11), severe corrosion of the reinforcing steel was evident, in spite of the fact that it had been cleaned of all corrosion product during the previous repair. In this case, the repair was of good quality in every detail except for the use of nonconductive material. Because of this oversight, the entire structure had to be demolished and replaced only a few years after the extensive repairs were made.

6.7   PROPERTIES OF COMMONLY USED REPAIR MATERIALS

There are many materials available for concrete repair. These materials include portland cement, latex-modified concrete, fibered cementitious systems, magnesium phosphate concrete, and a number of resinous systems. All of these materials perform satisfactorily under certain conditions. The key aspect to remember when selecting the material for concrete repair is its compatibility with the existing system and its performance requirements.

6.7.1   Cementitious Materials

Obviously, portland cement concrete or mortar are the best materials to use for repairs; however, shrinkage must be mitigated. It can be done through design of a mixture containing the minimal possible water content. The use of water-reducing admixtures is mandatory if a sufficiently low water/cement ratio is to be achieved.

Figure 6-10. Repaired beam that has been removed from structure for detailed investigation because of poor performance of epoxy mortar repair visible on end.

Figure 6-11. Beam shown in Figure 6-10 after epoxy mortar repairs were knocked off, showing extensive corrosion of reinforcing steel. The reinforcing was abrasive-blasted to white metal when repairs were made. It was conclusive of corrosion experts that new, more serious, corrosion resulted from changes of electrical potential due to mixing of conductive concrete with nonconductive epoxy mortar.

6.7.2 Latex-Modified Concrete and Mortar

Latex-modified cementitious compositions are similar to normal concrete or mortar, except they have a portion of the required water replaced with a latex solution. There are several different latex materials available; however, all look similar (milky white liquid) and perform about the same with one notable exception. Most of the formulations used with concrete are non-rewettable, that is, once they have skinned over, they will not rewet but rather act as bond breakers and will actually detract from bond rather than enhance it. This limitation requires very rapid mixing, placing, and finishing, usually within a period of 15 to 20 minutes, depending on the ambient temperature. Because of this limitation, special mixing and placing equipment is required for large placements; this equipment is not necessary for relatively small patching. A few formulations are available that are rewettable. They will soften with the application of moisture after their initial drying. While this property does have the advantage of allowing the work to be completed after drying, such materials obviously are not suitable where end use will result in continuous exposure to moisture.

Several manufacturers offer latex-modified mortars in kit form, usually a pail of the latex milk and a bag of the dry ingredients (cement, sand, and possible admixtures), which are mixed immediately prior to use. For large patches, gravel or other large aggregate can be added.

Latex-modified materials have several advantages that make them ideal for patching. Because the water content is minimized through use of the latex additive, shrinkage and thus bond problems are greatly reduced. Also, the latex material is itself an excellent bonding compound that provides additional adhesion at the bond line. Latex material requires only one day of wet curing and it is more resistant to water penetration, chemical attack, and abrasion damage than plain concrete.

6.7.3 Fibered Cementitious Systems

Either steel or plastic fibers can be added to a cementitious mixture. Their inclusion results in a random orientation throughout the mix, providing a continuous network that resists cracking and excessive shrinkage (which causes loss of the bond) of the patch. Fiber inclusion also increases the ductility (capability of being deformed without fracture) and impact resistance of the composition. Some manufacturers of latex-modified patching mortars now include fibers in their systems.

6.7.4   Magnesium Phosphate Concrete

Magnesium phosphate concrete is made by simply adding water to the prepared magnesium phosphate mixture, which is usually supplied in paper bags. It looks and acts similar to regular concrete; however, it is not a portland cement mixture but rather a synthetic concrete. It is very rapid curing and can reach a strength of about 2,000 psi in one hour. It is dimensionally stable, has high bond strength, and does not require wet curing. It is, thus, frequently used for emergency repairs in high-traffic areas and is one of the best materials for this purpose.

6.7.5   Resinous Systems

The physical properties of all resinous materials are very different from those of concrete, so their use should be restricted to those special applications where normal cementitious mixtures are not suitable. Whereas the stiffness of concrete is about the same regardless of temperature, most resins soften at elevated temperatures and become more brittle as the temperature is reduced. This category includes epoxy, polyester, and urethane.

Epoxy refers to a broad group of polymer formulations that can have a wide variety of final properties. To say "use epoxy" is not unlike specifying "use paint." Obviously, if a high-quality protective coating is required, one would not simply specify "paint." Likewise, it is not rational to simply specify "epoxy" without further qualification.

Epoxies can be formulated to be stiff, brittle, or flexible. They can set rapidly or with a long delay (hours or even days). They can also have many other properties. Of particular concern is the stiffness of these materials that changes with variation of temperature. Additionally, the coefficient of thermal expansion of most epoxy formulations is much different than that of concrete. For these reasons, if epoxy is used with concrete, apply it in the smallest volumes or thinnest sections possible to provide the desired result. Generally speaking, do not place epoxy or epoxy mortars in thicknesses greater than about ¼ in. or a maximum of ⅜ in.

Most epoxies are impermeable to air or gas and should not be used in applications that will prevent the concrete from breathing. Epoxy is often proposed for use to bond new concrete to old concrete. This use is not suitable except in those very rare instances where extremely high-strength concrete is involved. Since the bond strength of epoxy is many times the tensile strength of good-quality concrete, such strength is far more than required. Most of the time the use of such an expensive, difficult-to-work-with and trouble-prone compound cannot be justified.

Although epoxies require accurate proportioning, have to be mixed well, and are difficult to work with, there are many areas in which they are useful in the maintenance of concrete. As an adhesive, epoxies provide the strongest bond of any material commonly available. Therefore, epoxy is an ideal adhesive when a very strong bond is required. It has very high resistance to abrasion and chemical attack so it is used as a coating (or mixed with sand to form a mortar for thin overlays) when these qualities are required. Epoxy resins can be formulated to set and cure very rapidly, so sometimes they are used when fast turnaround repairs must be made.

When appropriately used, epoxy is one of the best materials available to the maintenance staff. However, it is not a magic solution for every possible problem, as it is often promoted, and should only be used after consideration of its suitability.

Polyesters can be formulated to appear much like epoxies, but they are subject to high shrinkage and do not offer the adhesive strength of epoxies. They are more resistant to chemical attack than concrete, but not to the same degree as epoxies. Polyester resins are, however, much less problem-prone and easier to use.

Urethanes are resinous materials that can be formulated as coatings, elastomers, chemical grouts, and foams. As coatings, urethanes are highly resistant to abrasion and chemical attack. As elastomers, they provide some of the highest-quality caulks and membranes available. Urethane chemical grouts are among the best materials for the stoppage of water infiltration through cracks or porous areas of concrete. Finally, as foams, they are among the best insulating materials available.

Urethanes are available in two distinct forms: single-component and two-component. The single-component systems require moisture from the atmosphere to cure, whereas the two-component systems are cured by a chemical reaction between the two components. Therefore, when used in relatively large masses, thick layers, or where the humidity is either very high or very low, the two-component systems are preferred.

There are a large number of other resinous materials on the market. Regardless of the exact generic type, be careful to always compare the physical properties of any proposed formulation to that of the concrete (or other material) with which it will be used. Because virtually all resinous formulations possess properties much different than concrete, in nearly all situations they should be used in the least thickness or quantity possible.

6.7.6 Sealers and Coatings

As water penetration is involved in nearly all forms of deterioration of concrete, using a coating or sealer to prevent water intrusion is often advised. However, because the concrete must be allowed to breathe, be sure that the particular covering material will be permeable to vapors. Both penetrating sealers, either clear or pigmented, and decorative coatings are available for concrete.

Most commonly used decorative coatings are permeable to vapors, as are well-known sealers such as silanes, siloxanes, and linseed soil. Many resinous materials such as most formulations of epoxy, polyester, and urethane are impermeable to vapor. Coatings impermeable to vapor should not be used in situations that will prevent moisture vapor movements within the concrete. It need not be a concern where only one side of an above-grade element is coated, but is an important consideration where complete encapsulation is the result, or for slabs on grade in locations subject to a high water table or freeze–thaw conditions.

6.8   REMOVALS AND SURFACE PREPARATION

When selectively removing concrete, it is important to limit removal methods to those that will not cause damage to the portions of the remaining structure. Likewise, when a concrete surface is prepared to receive new concrete or other material, preparation methods used should not unnecessarily damage or bruise the surface. Properly prepared surfaces should be free of broken or cracked aggregate, loose chips or fragments, and any microcracking. When making patches, the perimeter should be cut with a diamond saw to a depth of ¼ to ½ in. (Fig. 6-12). Doing so will provide straight lines for the patch and enough depth at the edge to assure the repair's durability. Even though some material manufacturers allow feather edging with their materials, it can result in early failures of otherwise good repairs.

6.8.1   Pneumatic and Electric Hammers

Frequently used tools for both removals and surface preparation are standard hand-held pneumatic or electric hammers. When they are used, only pointed gads or cutting tools should be allowed. Pointed tools are more likely to cut around large aggregate particles without damaging them. Because energy is limited to the small area of the point of the tool, they are, in general, less damaging than chistle-pointed tools or bushing hammers. Because standard hand-held pneumatic tools work on the principle of very high frequency and relatively small energy requirements per stroke, there is minimal chance of damage to adjoining concrete, regardless of the size or weight class of the tool. Note that hydraulically powered hand-held demolition tools are becoming

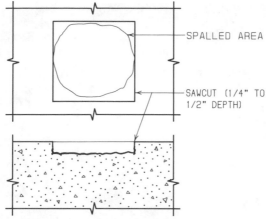

Figure 6-12. Typical sawcut around limits of patch.

increasingly popular due to their more efficient power transmission. While their operating principles are generally similar to those of pneumatic tools, hand-held hydraulically powered tools should be analyzed prior to use, due to their frequently greater power output.

Although hand-held tools will cause some minimal damage to the structure, increased use of so-called "hoe rams" or other equipment-mounted breakers that generally operate on the basis of relatively low frequency and high energy per blow can transmit damaging forces through the concrete section. Therefore, these tools should generally not be used for selective removals. When mounted breakers are used, the section of concrete to be removed should be separated from the remaining concrete by saw cutting (or other nondamaging methods) before final demolition with the breakers.

6.8.2   High-Pressure Water

High-pressure water is one of the best tools available to the maintenance engineer. High-pressure water can be used alone or can be combined with abrasive grit material. Its use ranges from simple surface cleaning to actual surface disintegration and concrete removal. When mixed with abrasive grit, it can even be used to cut through the concrete, including embedded reinforcements, much as an acetylene torch is used to cut through steel. Properly

Figure 6-13. Removal of concrete with high-pressure water (hydro-demolition).

applied, high-pressure water does not damage or bruise the surface (Hindo, 1987) and is an ideal preparation method. It is free of dust or serious vibration and is fairly quiet. The only real disadvantage is the requirement to furnish, handle, and dispose of the required water.

High-pressure water technology has made great strides within the last several years. In addition to large hydrodemolition machines, available today are relatively portable pump units that will generate pressures well in excess of 50,000 psi, with flow rates as low as 1 or 2 gal/min. Combined with a hand-held lance (Fig. 6-13) these units can be used to remove concrete from potholes and other similar defects. Other units are available to clean surfaces, remove coatings and sealers, or to scarify or remove the top ¼ to ½ in. or less of concrete.

6.8.3   Abrasive Blasting

Abrasive blasting, traditionally referred to as sandblasting, involves the propulsion of an abrasive material through a hose and nozzle onto a surface by means of compressed air, and has many uses in maintenance. Although ordinary sand has been the most frequently used abrasive, in special situations other types of more expensive abrasive materials have been employed. When using a fine, soft, round-grained sand, the procedure can simply clean a surface. By using a coarse, hard, angular sand, or special abrasive material, the surface concrete can be removed. Abrasive blasting can also be used for texturing, removing old coatings, and producing antiskid surfaces. In order to minimize the dust caused by this method, sometimes water is combined with the abrasive at the nozzle.

Although a great deal of work continues to be performed with traditional techniques, of particular interest to the maintenance engineer are the so-called dustless blasting methods. Such dustless methods involve impelling either abrasive material or metal shot onto the surface, then collecting it, along with any resulting cuttings, usually with a vacuum-type device. Dustless abrasive blasting can be performed with a hand-held nozzle with a special suction attachment connected to a vacuum unit with a flexible hose (Fig. 6-14). Of great usefulness are the dustless shot blasting machines (Fig. 6-15), of which a wide variety are available from several different manufacturers. All work on the same basic principle wherein the shot material is impelled onto the surface, usually by means of a centrifugal wheel. The shot material, along with the cuttings, are then recaptured by the machine where the cuttings are separated and removed by a vacuum unit, and shot material recirculated. The major advantages of the shot blasting machines are that the work can be accomplished free of dust or excessive noise, and the resulting surface is dry. An additional advantage of the dustless shot blasting machine is that it reuses the same shot material. This allows use of the optimal blasting material without excessive costs.

Other methods for both selective removal and surface preparation are sometimes used. Such alternate methodology is acceptable as long as the objective is realized, namely to provide a clean undamaged surface that is free of broken or cracked aggregate, loose chips, fragments, or microcracking.

6.9   REPAIR OF CONCRETE SLABS

Distress to floors and other slabs on grade nearly always manifests itself first at the joints. Joint-related distress is influenced largely by moving loads that cross the joints. Both the weight of the load and, perhaps more significantly, the size and composition of the wheels upon which the load is moved influence the degree of distress. The first distress noted by the maintenance person usually consists of spalling of the concrete adjacent to the joint, but spalling is very often not the cause but the result of distress. In these cases, the actual problem is pumping of the joint (vertical deflection of the concrete on one side relative to the other, as a load passes over), which is caused by slab curling.

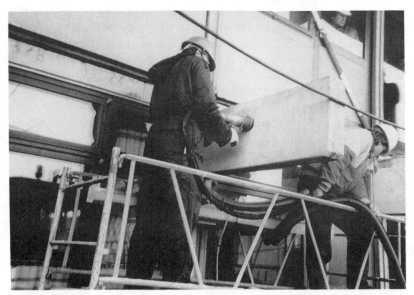

Figure 6-14. Dustless abrasive blasting with hand-held nozzle.

Figure 6-15. Surface preparation with dustless shot blasting machine.

Slab curling, shown in Figure 6-16, is caused when the top of the slab dries faster than the underside. This situation is very common in original construction, as the edges of the slab "curl" to a higher elevation than the rest of the slab. The condition can easily be detected by placing a straightedge over the joints. Often slab curling will go undetected for many years. Then, distress results from the differential movement of the slabs as loads cross over them. Such problems often appear shortly after new material-handling equipment (usually with smaller and harder tires) is used. In fact, many floor slabs that had performed well for several years have undergone joint distress solely because of changes in the material-handling equipment.

Where joint spalling is indicated, the maintenance engineer should ascertain whether joint pumping is occurring. This determination can be made by placing a finger over the joint and feeling for any movement as a load transverses the joint. If movement is detected, the joint must be stabilized prior to the spall repair, as distress will continue if the joints work under load.

Stabilization of pumping joints is accomplished by injecting a very thin cement grout under and into the joint. This practice is known as stitch grouting and involves drilling of small holes about 1½ in. in diameter through the slab, immediately adjacent to the joint. The holes are spaced about 1½–2 ft apart and are staggered on each side of the joint (Fig. 6-17). Because the slab surface at the joint is already higher than the main portion of the slab (the result of slab curling), exercise great care not to raise the slab during the grout injection.

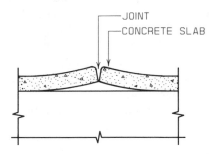

Figure 6-16. Exaggerated view of typical slab curling at joint.

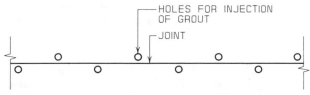

Figure 6-17. Hole layout for stitch grouting to correct pumping of joints due to slab curling.

Spalling at the joints can also be caused by impact of wheels or other objects against the joint edge, which is often not supported. Where steel or other small hard-wheeled equipment is operated, the joints should be filled with a semirigid filler. Semirigid epoxy joint fillers formulated especially for this purpose are available. Care must be taken not to use just any epoxy since most achieve high bond strength, a quality that must be avoided in joint filling. The objective is to fill the joint with a rigid enough material to prevent impact damage, not to "glue" the joint together. Isolated spalls should be prepared as previously outlined and filled with a suitable material. As previously discussed, cementitious materials are preferred as their physical properties are similar to concrete.

In order to minimize shrinkage, spalls should be filled with a stiff-consistency material, rammed densely into place. Whereas normal cement mortars or concrete can be effectively used for spall repair, the latex-modified cementitious systems are especially suited for repairs. These materials result in greater bond and less shrinkage. Where they are used, prime the spall surface with a bond coat, such as a mixture of approximately equal parts of latex milk, cement, and sand. As with plain mortar or concrete, the latex mortar should be placed in as stiff a consistency as can be handled.

Where extensive spalling has occurred at a joint, repairs can be made as follows:

• For damaged widths less than 1½ in., sawcut two straight lines to a depth of at least 1 in., as shown in Figure 6-18. Remove the concrete and fill the resulting slot with simirigid epoxy, formulated especially for this use. An improperly selected epoxy material will bond the two sides of the joint together, which must be avoided.

• For damaged widths from 1½ in. to about 8 in., sawcut two straight lines, remove the concrete, and fill with a silica-filled semirigid epoxy (as above), except provide a divider strip of plastic (or other similar material) centered over the joint as indicated in Figure 6-19. If the joint is open more than about ⅛ in., fill with sand or restrict flow with a suitable sealer such that the epoxy will not leak into it.

• For damaged widths greater than about 8 in., sawcut two lines, remove concrete full depth and replace it with new concrete. Steel dowels or a bonding agent should be provided on one side of the cut to tie the new concrete to the original. The opposite side should be coated with a bond breaker to allow the finished joint to work (Fig. 6-20).

A variety of chemical sealers and coatings are available for concrete floors. Remember that concrete must be allowed to breathe, so any material used to coat a slab on grade should be permeable to vapor. Floors suffering from excessive wear so that the surface layer has abraded will probably require a

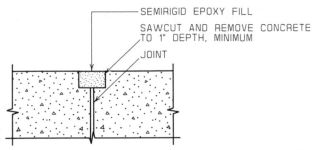

Figure 6-18. Typical correction of joint spalls with maximum width of about 1½ in.

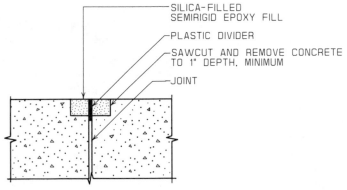

Figure 6-19. Typical correction of joint spalls with maximum width of about 8 in.

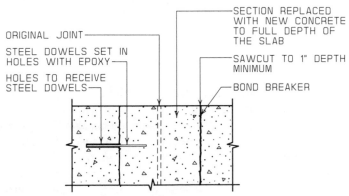

Figure 6-20. Typical correction of joint spalling with width greater than about 8 in.

partial or complete overlay. The overlay may be of either a resinous or cementitious nature depending on the exact requirements of the floor. Such extensive work goes beyond ordinary maintenance and will require the services of a specialty repair contractor. Where such extensive wear to a floor has occurred, often housekeeping has been lacking. Keeping floors free of debris is an important aspect in maintenance, especially where forklift traffic exists.

6.10 STRUCTURAL REPAIRS

Most structural repairs go far beyond ordinary maintenance; however, it is imperative that the maintenance staff be constantly alert to potential deficiencies that could lead to more extensive repair requirements. Concrete structures should receive regular visual inspections. Any questionable conditions should be given immediate attention. Rust stains, even minor ones, could be an early warning of greater potential problems. To ascertain the cause of the stains usually requires removal of the concrete to the area of the stains' origin. The stains could be the result of corrosion of an unintentionally embedded ferrous metal object such as a nail, form tie, or tool. Should this be the case, the metal object should be removed and the removed concrete replaced. Either plain portland cement mortar or concrete or latex-modified patching materials may be used. If it is found that the origin of the corrosion is the reinforcing steel or other intentionally embedded metal, a potentially serious problem exists. It should be promptly evaluated by a qualified professional.

All concrete cracks, so some cracks should be expected in any concrete structure. The maintenance staff should be alert for new cracks developing, old cracks that are noticeably opening, and cracks that are exuding any foreign material, especially rust-colored substances. Such conditions indicate a more serious condition, and again should be investigated by a qualified individual.

Nonmoving cracks are usually best left alone unless they create unacceptable conditions such as water leakage or unacceptable visual appearance. When water leakage is the problem, the crack can be routed out and filled with an elastomer. It is important to provide a bond-breaker at the bottom of the slot (Fig. 6-21) to prevent restraint, precluding the elastomer to expand. Subsurface cracks that are leaking can be remedied by pressure injection of a chemical grout material. However, this correction is usually performed by a specialist contractor, and is beyond the resources of most maintenance personnel.

It is nearly impossible to repair cracks so they are not visible. If they must be hidden for aesthetic reasons, decorative coating over the repaired surface will probably be necessary.

There has been a tendency in recent years to fill all cracks with epoxy. This

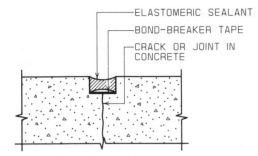

Figure 6-21. Typical installation of bond-breaker tape under new elastomer.

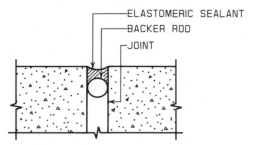

Figure 6-22. Typical installation of backer rod in deep joints in order to maintain proper elastomer configuration.

practice is quite expensive, can result in further damage if the cracks are working (a condition difficult to ascertain with only visual observation), and is impossible to repair properly if the cracks are contaminated with dirt or debris. Working cracks injected with epoxy cannot move, so new cracks will form unless other joints to accommodate movement are provided.

Movement joints are often provided during original construction. These joints are generally filled with an elastomeric material that is designed to withstand the anticipated movement. These joints should be regularly inspected and maintained in proper condition. Where the elastomeric filler material has deteriorated or lost adhesion to the substrate, it must be removed and replaced. It is important to maintain the proper shape of such joints to ensure performance. Generally speaking, the width of the elastomer should be approximately twice the depth. Where a deep joint is involved, joint backer rod (Fig. 6-22) should be installed to maintain the correct configuration. With random cracks or tight joints, rout out the joint and provide it with a bond-breaker tape as indicated on Figure 6-21. These are very important details, and joints not made accordingly will suffer early distress.

6.11   SUMMARY

Concrete is one of the most durable and versatile materials available for use in building construction. It is made up of cement, aggregate, and water. Water is used for hydration of cement and for obtaining the plasticity necessary for placement of concrete. If too much water is used, the concrete will be porous, it will tend to absorb water, and it will lack durability. For obtaining durable concrete, water content should be maintained at a minimum level. This rule applies to concrete in new construction as well as to the concrete used in repairs.

It is imperative that materials used for repairs are compatible with the existing concrete. Dimensional stability, modulus of elasticity, and coefficient of thermal expansion of concrete repair materials must be compatible with the concrete that is being repaired. For best results, concrete surfaces should be designed and constructed to drain. The susceptibility of porous concrete to absorb water is intensified by the ponding of water.

Providing sufficient amount of cover over reinforcing steel is another important factor in concrete construction and concrete repair. Proper design and construction of expansion joints to allow for controlled movement of concrete elements is also crucial.

When considering overlays and repairs of spalled areas on concrete surfaces, surface preparation of concrete is of paramount importance. The quality and durability of concrete repairs depends, to a great extent, on how the surfaces of existing concrete were prepared.

To ensure that concrete surfaces perform for the expected periods of time, it is critical that inspections, maintenance, and repairs are performed on a regular basis. Often, major costs of concrete repairs can be averted if minor concrete failures are repaired when they first appear.

6.12   SUGGESTED READINGS

Building Code Requirements for Reinforced Concrete (ACI 318-83). 1983. Detroit: American Concrete Institute.

Concrete Manual, 8th ed. 1981. Washington, DC: Bureau of Reclamation, U.S. Department of the Interior.

Design and Control of Concrete Mixtures, 12th ed. 1977. Skokie, IL: Portland Cement Association.

Hindo, Kal R. 1990. In-place testing and surface preparation of concrete. *Concrete International* 12(4, April).

Warner, James. 1984. Important aspects of cementitious materials used in repair and retrofit. In *Proceedings of the Eighth World Conference on Earthquake Engineering*. San Francisco.

6.13   ASSOCIATIONS WITH INFORMATION ON CONCRETE CONSTRUCTION

American Concrete Institute
P.O. Box 19150 Redford Station
Detroit, MI 48219-0150

International Association of Concrete Repair Specialists
P.O. Box 17402
Dulles International Airport
Washington, DC 20041

Portland Cement Association
5420 Old Orchard Road
Skokie, IL 60077-1030

World of Concrete
426 South Westgate
Addison, IL 60101

7. Exterior Wood Surfaces

William C. Feist

7.1 INTRODUCTION

The primary functions of any wood finish (e.g., paint, varnish, wax, stain, oil) are to protect the wood surface, help maintain appearance, and provide cleanability. Unfinished wood can be used outdoors without protection. However, wood surfaces exposed to the weather without any finish are roughened by photodegradation and surface checking, change color, and slowly erode.

Wood and wood-based products in a variety of species, grain patterns, textures, and colors can be effectively finished or refinished by many different methods. Selection of the finish will depend on the appearance and degree of protection desired and on the substrates used. Also, different finishes give varying degrees of protection; therefore, the type, quality, quantity, and application method of the finish must be considered when selecting and planning the finishing or refinishing of wood and wood-based products.

7.2 WOOD PROPERTIES

Understanding wood properties is integral to effective performance of exterior finishes and reducing maintenance. Wood surfaces that shrink and swell the least are best for painting. For this reason, vertical or edge-grained surfaces (Fig. 7-1) are far better than flat-grained surfaces of any species, especially for exterior use where wide ranges of relative humidity and periodic wetting can produce equally wide ranges of swelling and shrinking. Also, because the swelling of wood is directly proportional to density, low-density species are preferred over high-density species. Softwoods (cedar, redwood, pine) are generally better than hardwoods (oak, birch, beech) for painting. However, even high-swelling and dense wood surfaces with a flat grain can be stabilized with a resin-treated paper overlay (overlaid exterior plywood and lumber) to

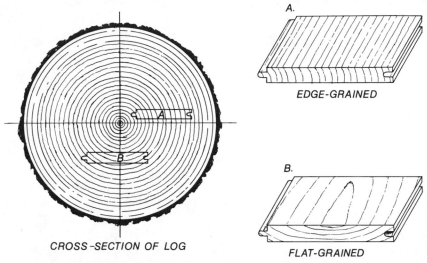

A.

EDGE-GRAINED

B.

FLAT-GRAINED

CROSS-SECTION OF LOG

Figure 7-1. Effect of sawing method on ring orientation in lumber.

Figure 7-2. Paint applied over edge-grained boards (top and bottom). It performs better than that applied to flat-grained boards (middle).

provide excellent surfaces for painting. Medium-density, stabilized fiber-board products with a uniform, low-density surface or paper overlay are also good substrates for exterior use. However, edge-grained heartwood of western redcedar and redwood are the species most widely used for exterior siding and trim when painting is desired. These species are classified as those easiest to keep painted.

Edge-grained surfaces of all species are considered excellent for painting, but most species are available only as flat-grained lumber. The flat-grained woods are classified as good to fair for paint-holding characteristics (Fig. 7-2). Many flat-grained species are commonly painted, particularly the pines, Douglas fir, and spruce, but these species usually require more care and attention than the edge-grained surfaces. Flat-grained boards that are to be painted should be installed in areas protected from rain and sun.

Three general categories of wood products are commonly used in exterior construction: (1) lumber, (2) plywood, and (3) reconstituted wood products such as hardboard (a fiberboard) and particleboard. Each has characteristics that will affect the durability of any finish applied to it.

7.2.1   Lumber

Lumber is being used less and less as an exterior siding, but it was once the most common wood material used in construction. Many older buildings have solid wood siding so maintaining the lumber is an important consideration in respect to maintenance of older structures. Today, a large amount of solid lumber is used for decking, flooring, fences, and similar items. The ability of lumber to absorb and retain a finish is affected by species, by ring direction with respect to the surface (or how the piece was sawn from the log), and by smoothness.

The weight of wood varies tremendously between species. Some common construction woods, such as southern pine, are dense and heavy compared with lighter-weight woods, such as redwood and cedar. The weight of wood is important because heavy woods shrink and swell more than light ones. This dimensional change in lumber occurs as the wood gains or loses moisture. Excessive dimensional change in wood puts constant stress on the paint film and may result in early failure of the finish.

Ring direction also affects paint-holding characteristics and is determined at the time lumber is cut from a log (Fig. 7-2). Most standard grades of lumber contain a high percentage of flat grain. Lumber used for board and batten siding, drop siding, or shiplap is frequently flat-grained. Bevel siding is commonly produced in several grades. In some cases, the highest grade of lumber is required to be edge-grained and all heartwood over most of the width for greater paint durability. Other grades may be flat-grained, edge-grained, or mixed-grain and without requirements as to heartwood. Some species have wide bands of earlywood and latewood (Fig. 7-3). Wide, prominent bands of latewood are characteristic of southern pine and most Douglas fir, and paint will not hold well on these species.

7.2.2   Plywood

Exterior plywood with a rough-sawn surface is commonly used for siding. Smooth-sanded plywood is not recommended for siding, but it is often used in soffits. Plywood surfaces are predominantly flat-grained. Both sanded and rough-sawn plywood will develop surface checks (called face-checking), especially when exposed to moisture and sunlight. These surface checks can lead to early paint failure when using oil or alkyd paints (Fig. 7-4). Quality stain-blocking acrylic latex primers and topcoat paints generally perform better. The flat-grained pattern present in nearly all plywood can also contribute to early paint failure. Therefore, if smooth or rough-sawn plywood is to be painted or repainted, special precautions should be exercised. Semitransparent, penetrating, oil-based stains are often more appropriate for rough-sawn exterior plywood surfaces, but quality acrylic latex paints also perform very well.

7.2.3   Reconstituted Wood Products

Reconstituted wood products are made by forming small pieces of wood into large sheets, usually 4 × 8 ft or as required for a specialized use, such as beveled siding. These products may be classified as fiberboard or particleboard, depending on the nature of the basic wood component. Fiberboards are produced from mechanical pulps. Hardboard is a relatively heavy type of fiberboard, and its tempered or treated form, designed for outdoor exposure, is used for exterior siding. Hardboard siding is usually factory-primed. It is often sold in 4- × 8-ft sheets as a substitute for solid wood-beveled siding. Particleboards are manufactured from whole wood in the form of splinters, chips, flakes, strands, or shavings. Waferboard, oriented strandboard, and flakeboard are three types of particleboard made from relatively large flakes or shavings. Some reconstituted wood products may be factory-primed with paint, and some may even have a factory-applied topcoat. Also, some may be overlaid with a resin-treated cellulose fiber sheet to provide a superior surface for paint.

7.2.4   Preservative-Treated Wood

Sometimes wood is used in severe outdoor situations where special treatments and finishes such as preservatives are required for proper protection and the

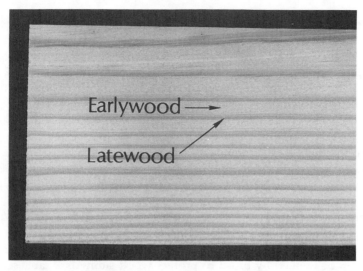

Figure 7-3. Earlywood and latewood bands in southern yellow pine. These distinct bands often lead to early paint failure. Therefore, penetrating stains are preferred.

Figure 7-4. Early paint failure on plywood due to penetration of moisture into surface checks.

best service. These situations involve the need for protection against decay (rot) and insects.

There are two main types of preservatives: (1) preservative oils (e.g., coal tar creosote, organic solvent solutions of pentachlorophenol), and (2) water-borne salts (e.g., chromated copper arsenate [CCA]). These preservatives can be applied in several ways, but pressure treatment always gives the greatest protection against decay. Higher preservative content of pressure-treated wood generally results in greater resistance to weathering and improved surface durability. The chromium-containing preservatives also protect against ultraviolet degradation, an important factor in the weathering process. Some preservative treatments protect against decay and insects and others provide insect repellency.

Water-repellent preservatives are introduced into wood by a vacuum/pressure or dipping process (National Wood Window and Door Manufacturer's Association Industry Standard 4-81). These pretreatments can be painted on. They can be applied by brush to protect wood siding (see later section on paints). Coal tar creosote or other heavy-oil preservatives tend to stain through paint, especially light-colored paint, unless the treated wood has weathered for many years before painting.

7.2.5   Fire-Retardant Coated Wood

Many commercial fire-retardant coating products are available that provide varying degrees of protection for wood against fire. These paint coatings generally have low surface flammability characteristics and intumesce (swell) to form an expanded low-density film upon exposure to fire, thus insulating the wood surface below from heat and retarding pyrolysis reactions. The paints have added ingredients to restrict the flaming of any released combustible vapors. Chemicals may also be present in these paints to promote decomposition of the wood surface to charcoal and water, rather than forming volatile flammable products.

Most fire-retardant coatings are intended for interior use, but some are available for exterior application. Conventional paints have been applied over the fire-retardant coatings to improve their durability. Most conventional decorative coatings will in themselves slightly reduce the flammability of wood products when applied in conventional film thicknesses.

In addition to fire-retardant coatings, wood may be pressure-treated with fire retardants. A number of different systems are used. They are divided into two types: interior and exterior. The fire-retardant treatment of wood does not generally interfere with adhesion of decorative paint coatings, unless the treated wood has an extremely high moisture content because of its increased hygroscopicity (absorption of water from the atmosphere). It is most important that only those fire-retardant treatments specifically prepared and recommended for outdoor exposure be used for that purpose. These treated woods are generally painted according to recommendations of the manufacturer rather than being left unfinished because the treatment and subsequent drying often darken and irregularly stain the wood.

7.3   NATURAL WEATHERING OF WOOD

Wood exposed outdoors undergoes a number of physical and chemical phenomena mostly caused by moisture, sunlight, and temperature. Being a product of nature, wood is also subject to biological attack by fungi and insects.

The simplest finish for wood is created by the natural weathering process. Without paint or treatment of any kind, wood surfaces gradually change in color and texture and then stay almost unaltered for a long time, if the wood does not decay (rot). Generally, the dark-colored woods become lighter and the light-colored woods become darker. As weathering continues, all woods become gray, accompanied by photodegradation and gradual loss of wood cells at the surface. As a result, exposed unfinished wood will slowly wear away in a process called erosion.

The weathering process is a surface phenomenon and is so slow that most softwoods (redwood, pine, Douglas-fir) wear away (erode) at an average rate of about ¼ in. per century. Dense hardwoods (oak, ash) erode at a rate of only ⅛ in. per century. Very low-density softwoods, such as western redcedar, may erode at a rate as high as ½ in. per century. In cold northern climates, erosion rates as low as ¹⁄₃₂ in. per century have been reported for pine. Accompanying this erosion or loss of wood substance on the wood surface are the swelling and shrinking stresses caused by fluctuations in moisture content. These stresses result in surface roughening, grain raising, differential swelling of earlywood and latewood bands, and the formation of many small parallel checks and cracks. Larger and deeper cracks may also develop and warping frequently occurs (Fig. 7-5).

The weathering process is usually accompanied by the growth of dark-colored spores and mycelia of fungi or mildew on the surface, which gives the wood a dark gray, uneven, and unsightly appearance. In addition, highly colored wood extractives in such species as western redcedar and redwood add to the variable color of weathered wood. The dark brown color of extractives may persist for a long time in areas not exposed to the sun and where the extractives are not removed by the washing action of rain.

7.4   APPLIED FINISHES TO EXTERIOR WOOD SURFACES

A number of different finishes can be used to protect outdoor-exposed wood. Different degrees of protection will be provided to the wood surface from the

Figure 7-5. Weathered surface of softwood after 15 years of exposure in Madison, Wisconsin.

different types of finishes. Paints, solid-color stains, semitransparent stains, water-repellent preservatives, and water repellents and transparent coatings are the most commonly used finishes. The various types of wood preservatives and finishes for exterior wood and their maintenance are summarized in Table 7-1; finishing methods and suitability are summarized in Table 7-2.

7.4.1   Paints

Paints are coatings commonly used on wood and provide the most protection because they block the damaging ultraviolet light rays from the sun. They are available in a wide range of colors and may be either oil-based or latex-based. Latex-based paints are waterborne and oil-based (also called alkyd) paints are organic solventborne. Paints are used for esthetic purposes, to protect the wood surface from weathering, and to conceal certain defects.

Paints are applied to the wood surface and do not penetrate it to any great extent. The wood grain is completely obscured and a surface film is formed. Paints perform best on smooth, lightweight, edge-grained lumber. This surface film can blister or peel if the wood becomes wet or if inside water vapor moves through the house wall and into the wood siding because of the absence of a vapor-retarding material. Latex paints are porous and will allow some moisture movement. Some oil-based paints are resistant to moisture movement.

Of all the finishes, paints provide the most protection for wood against surface erosion and offer the widest selection of colors. A nonporous paint film retards penetration of moisture and reduces the problem of discoloration by wood extractives, paint peeling, and checking and warping of the wood.

It is very important to note that paint is not a preservative. It will not prevent decay if conditions are favorable for fungal growth. Original and maintenance costs are often higher for a paint finish than for a water-repellent preservative or a penetrating stain finish.

7.4.2   Solid-Color Stains

Solid-color stains (also called opaque, hiding, or heavy-bodied) are pigmented finishes similar to paint. They are available in a wide range of colors and are made with a much higher concentration of pigment than the semitransparent penetrating stains. Solid-color stains will totally obscure the natural color and grain of wood. Oil-based (alkyd) solid-color stains form a film much like paint and as a result can peel loose from the substrate. Latex-based solid-color stains also form a film. Both these stains are similar to thinned paints and can usually be applied over old paint or stains if the old finish is clean and securely bonded to the wood. Application of two coats will give much better performance than one coat.

7.4.3   Semitransparent Penetrating Stains

Semitransparent oil-based penetrating stains are only moderately pigmented and do not totally hide the wood grain. These stains penetrate the wood surface, are porous to water vapor, and do not form a surface film like paints. As a result, they will not blister or peel even if moisture moves through the wood. Penetrating stains are alkyd or oil-based, and some may contain a fungicide (preservative or mildewcide) as well as a water repellent. Moderately pigmented semitransparent latex-based (waterborne) stains are also available, but they do not penetrate the wood surface as do the oil-based stains.

The semitransparent stains are most effective on rough-sawn lumber or plywood surfaces. They will not perform as well on smooth surfaces, but their performance will improve after recoating. These stains are available in a variety of colors and are especially popular in the brown or red earth tones

Table 7-1. Exterior Wood Finishes: Types, Treatment, and Maintenance[a]

Finish	Initial Treatment	Appearance of Wood	Cost of Initial Treatment	Maintenance Procedure	Maintenance Period of Surface Finish	Maintenance Cost
Preservative oils (creosotes)	Pressure, hot and cold tank steeping	Grain visible; brown to black in color, fading slightly with age	Medium	Brush down to remove surface dirt	5–10 years only if original color is to be renewed; otherwise no maintenance is required	Nil to low
Waterborne preservatives	Brushing	Grain visible; brown to black in color, fading slightly with age	Low	Brush down to remove surface dirt	3–5 years	Low
	Pressure	Grain visible; greenish or brownish in color, fading with age	Medium	Brush down to remove surface dirt	None, unless stained, painted, or varnished as below	Nil, unless stains, varnishes, or paints are used as below
	Diffusion plus paint	Grain and natural color obscured	Low to medium	Clean and repaint	7–10 years	Medium
Organic solvent preservatives[b]	Pressure, steeping, dipping, brushing	Grain visible; colored as desired	Low to medium	Brush down and reapply	2–3 years or when preferred	Medium

192

Finish	Appearance of wood		Maintenance procedure	Maintenance period	
Water repellent[c]	One or two brush coats of clear material or, preferably, dip-applied	Grain and natural color visible, becoming darker and rougher textured	Low	Clean and apply sufficient finish	1–3 years or when preferred	Low to medium
Semitransparent stains	One or two brush coats	Grain visible; color as desired	Low to medium	Clean and apply sufficient finish	3–6 years or when preferred	Low to medium
Clear varnish	Three coats (minimum)	Grain and natural color unchanged if adequately maintained	High	Clean and stain bleach areas; apply two more coats	2 years or when breakdown begins	High
Paint	Water-repellent, prime, and two topcoats	Grain and natural color obscured	Medium to high	Clean and apply topcoat, or remove and repeat initial treatment if damaged	7–10 years[d]	Medium

Source: *Wood Handbook: Wood as an Engineering Material*, Agricultural Handbook No. 72. 1987. Madison, WI: Forest Products Laboratory, USDA Forest Service.

[a]This table is a compilation of data from the observation of many researchers.

[b]Pentachlorophenol, bis (tri-n-butyltin oxide), copper naphthenate, copper-8-quinolinolate, and similar materials.

[c]With or without added preservatives. Addition of preservative helps control mildew growth.

[d]Using top-quality acrylic latex topcoats.

Table 7-2. Finishing Methods for Exterior Wood Surfaces: Suitability[a]

Type of Exterior Wood Surfaces	Water-repellent Preservative		Semitransparent Stains		Paints	
	Suitability	Expected Life[b] (yrs)	Suitability	Expected Life[c] (yrs)	Suitability	Expected Life[d] (yrs)
Siding						
Cedar and redwood						
Smooth (vertical grain)	High	1–2	Moderate	2–4	High	4–6
Roughsawn or weathered	High	2–3	Excellent	5–8	Moderate	3–5
Pine, fir, spruce, etc.						
Smooth (flat-grained)	High	1–2	Low	2–3	Moderate	3–5
Rough (flat-grained)	High	2–3	High	4–7	Moderate	3–5
Shingles						
Sawn	High	2–3	Excellent	4–8	Moderate	3–5
Split	High	1–2	Excellent	4–8		
Plywood (Douglas fir and southern pine)						
Sanded	Low	1–2	Moderate	2–4	Moderate	3–5
Textured (smooth)	Low	1–2	Moderate	2–4	Moderate	3–5
Textured (roughsawn)	Low	2–3	High	4–8	Moderate	4–6
Medium-density overlay[e]			Excellent		Excellent	6–8
Plywood (cedar and redwood)						
Sanded	Low	1–2	Moderate	2–4	Moderate	3–5
Textured (smooth)	Low	1–2	Moderate	2–4	Moderate	3–5
Textured (roughsawn)	Low	2–3	Excellent	5–8	Moderate	4–6

	Appearance	Years	Appearance	Years	Appearance	Years
Hardboard, medium density[e][f]						
Smooth						
Unfinished					High	4–6
Preprimed					High	4–6
Textured						
Unfinished					High	4–6
Preprimed					High	4–6
Millwork (usually pine)						
Windows, shutters, doors, exterior trim	High[g]		Moderate	2–3	High	3–6
Decking						
New (smooth)	High	1–2	Moderate	2–3	Low	2–3
Weathered (rough)	High	2–3	High	3–6	Low	2–3
Glued-laminated members						
Smooth	High	1–2	Moderate	3–4	Moderate	3–4
Rough	High	2–3	High	6–8	Moderate	3–4
Waferboard			Low	1–3	Moderate	2–4

Source: Wood Handbook: Wood as an Engineering Material, Agricultural Handbook No. 72. 1987. Madison, WI: Forest Products Laboratory, USDA Forest Service.

[a]These data were compiled from the observations of many researchers. Expected life predictions are for an average location in the continental United States; expected life will vary in extreme climates or exposure (desert, seashore, deep woods, etc.).

[b]Development of mildew on the surface indicates a need for refinishing.

[c]Smooth, unweathered surfaces are generally finished with only one coat of stain, but rough-sawn or weathered surfaces, being more adsorptive, can be finished with two coats, with the second coat applied while the first coat is still wet.

[d]Expected life of two coats, one primer and one topcoat. Applying a second topcoat (three-coat job) will approximately double the life. Top-quality acrylic latex paints will have the best durability.

[e]Medium-density overlay is generally painted.

[f]Semitransparent stains are not suitable for hardboard. Solid-color stains (acrylic latex) will perform like paints. Paints are preferred.

[g]Exterior millwork, such as windows, should be factory-treated according to Industry Standard IS4–81. Other trim should be liberally treated by brushing before painting.

because they give a natural or rustic appearance. They are an excellent finish for old weathered wood. Semitransparent, penetrating, oil-based stains are not effective when applied over a solid-color stain or over old paint. They are not recommended for use on hardboard, flakeboard, or similar panel products.

7.4.4 Water-Repellent Preservatives and Water Repellents

A water-repellent preservative may be used as a natural finish. It contains a fungicide or mildewcide (the preservative), a small amount of wax for water repellency, a resin or drying oil, and a solvent, such as turpentine or mineral spirits. Water-repellent preservatives do not contain coloring pigments. Therefore, the resulting finish will vary in color depending on the species of wood. The mildewcide also prevents wood from darkening (graying). The initial application of the water-repellent preservative to smooth wood surfaces may last only a year or two. The finish will have to be applied regularly. Subsequent treatments are usually more durable and may last 3 to 4 years. After several refinishings, the wood may need refinishing only when the surface starts to become unevenly colored by fungi.

Water-repellent preservatives may also be used as a treatment for bare wood before priming and painting or in areas where old paint has peeled, exposing bare wood, particularly around butt joints or in corners (Figs. 7-6 and 7-7). This treatment keeps rain or dew from penetrating into the wood,

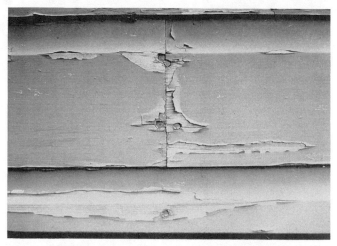

Figure 7-6. Paint normally fails first around the ends and edges of a board. Liberal application of a water repellant or water-repellent preservative especially to the end grain can prolong paint life in these areas.

Figure 7-7. Window sash and frame treated with a water-repellent preservative and then painted (*top*), and window sash and frame not treated before painting (*bottom*). Both treatments were weathered for 5 years. Note the normally weathered paint, good condition of the wood, and glazing on the treated structure.

especially at joints and end grain, and thus decreases the shrinking and swelling of wood. As a result, less stress is placed on the paint film, and its service life is extended. Water-repellent preservatives are also used as edge treatments for panel products like plywood.

Water repellents are also available. These are water-repellent preservatives with the fungicide, mildewcide, or preservative left out. Water repellents are not effective natural finishes by themselves but can be used as a stabilizing treatment before priming and painting. Water repellents are often used to protect decks and fences made from pressure-treated wood with waterborne preservatives.

7.4.5 Transparent Coatings

Clear coatings of conventional spar or marine varnishes, which are film-forming finishes, are not generally recommended for exterior use on wood. Shellac or lacquers should never be used outdoors because they are very sensitive to water and are very brittle. Varnish coatings become brittle by exposure to sunlight and develop severe cracking and peeling, often in less than 2 years. Refinishing will often involve removing all the old varnish. Areas that are protected from direct sunlight by overhang, or are on the north side of the structure, can be finished with exterior-grade varnishes. Even in protected areas, a minimum of three coats of varnish is recommended, and the wood should be treated with a water-repellent preservative before finishing. Using compatible pigmented stains and sealers as undercoats will also contribute to a greater service life of the clear varnish finish. In marine exposures, six coats of varnish should be applied for best performance.

7.5 FINISH FAILURES

Painted wood appears somewhere on practically every house and on most commercial buildings. When properly applied to the appropriate type of wood substrate, paint can give a service life of up to 10 years. All too often, however, problems may develop during the application of the paint or the paint coat fails to achieve the expected service life.

Paint properly applied and exposed under normal conditions is usually not affected by the first 2 to 3 years of exposure. Areas that deteriorate the fastest are those exposed to the greatest amount of sun and rain, usually on the south and west sides of a building. The normal deterioration process leads first to soiling or accumulation of dirt and then to a flattening stage when the coating gradually starts to chalk and erode away. However, early paint failure may develop under certain conditions of service. Excess moisture, flat-grained wood, high porosity in the coating, and applying a new paint coat without proper preparation of the old surface can all contribute to early paint failure. Mildew (fungus growth), extractives in wood, rust, and knot stains, and improper application of coatings may contribute to the failure of finishes.

The most common cause of premature finish failure on wood is moisture. Finishes on the outside walls of houses are subject to wetting from rain, dew, and frost. Equally serious is "unseen" moisture moving from inside the house to the outside, which is particularly true on buildings without a proper vapor-retarding material in cold northern climates.

7.5.1   Temperature Blisters

Temperature blisters are bubblelike swellings that occur on the surface of the paint film as early as a few hours after painting or as long as one to two days later (Fig. 7-8). They occur only in the last coat of paint. They are caused when a thin dry skin has formed on the outer surface of the fresh paint and the liquid thinner in the wet paint under the dry skin changes to vapor and cannot escape. A rapid rise in temperature, as when the rays of the sun fall directly on freshly painted surfaces, will cause the vapors to expand and produce blisters. Usually only oil-based paint blisters in this way. Dark colors that absorb heat and thick paint coats are more likely to blister than white paints or thin coats.

7.5.2   Moisture Blisters

Moisture blisters are also bubblelike swellings of the paint film on the wood surface (Fig. 7-9). As the name implies, they usually contain moisture when they are formed. They may occur where outside moisture such as rain enters the wood through joints and other end-grain areas of boards and siding. Moisture may also enter from poor construction and maintenance practices. Damage appears after spring rains and throughout the summer. Paint failure is most severe on the sides of buildings facing prevailing winds and rain. Blisters may occur in both heated and unheated buildings.

Moisture blisters may also result from inside water moving to the outside. Plumbing leaks, overflow of sinks, bathtubs or shower spray, and improperly sealed walls are sources of inside water. Such damage is not seasonal and occurs when the faulty condition develops.

Figure 7-8. Temperature blisters can result when partially dried paint is suddenly heated by the direct rays of the sun.

Figure 7-9. Moisture blisters can result from water vapor moving to the outside from within the house.

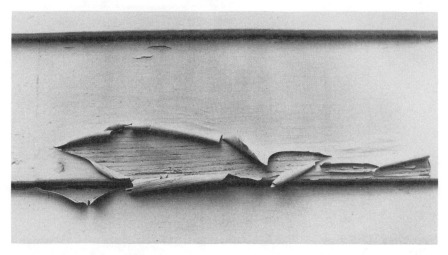

Figure 7-10. Intercoat peeling of paint is usually caused by poor preparation of the old surface.

7.5.3 Intercoat Peeling

Intercoat peeling (Fig. 7-10) is the separation of the new paint film from the old paint coat, indicating a weak bond between the two. Intercoat peeling usually results from inadequate cleaning of the weathered paint and usually occurs within 1 year of repainting. This type of paint peeling can be prevented by following good painting practices.

7.5.4 Cross-Grain Cracking

Cross-grain cracking (Fig. 7-11) occurs when oil-based or alkyd paint coatings become too thick. This problem often occurs on older homes that have been painted many times. Paint usually cracks in the direction it was brushed onto the wood. Cross-grain cracks run across the grain of the paint. Once cross-grain cracking has occurred the only solution is to completely remove the old paint and apply a new finishing system on the bare wood.

7.5.5 Chalking

Chalking (Fig. 7-12) is a property of the paint and results when the paint film gradually weathers or deteriorates, releasing the individual particles of pigment. These individual particles act like a fine powder on the paint surface. Most

Figure 7-11. Cross-grain cracking results from an excessive buildup of paint.

Figure 7-12. Some paints or stains chalk badly and can discolor a lower surface as they wash down over it.

paints chalk to some extent. Some nonchalking paints may be available. Some chalking may be desirable since it allows the paint surface to be self-cleaning. However, chalking is objectionable when it washes over a surface with a different color or when it causes premature disappearance of the paint film through excess erosion.

7.5.6 Mildew

Mildew (Fig. 7-13) is probably the most common cause of house paint discoloration and gray discoloration of unfinished wood. Mildew is a form of stain fungi or microscopic plant life. The most common species are black, but some are red, green, or other colors. It grows most extensively in warm, humid climates but is also found in cold northern states. Mildew may be found anywhere on a building, but it is most common on walls behind trees or shrubs where air movement is restricted. Mildew may also be associated with the dew pattern of the house. Dew will form on those parts of the house that are not heated and tend to cool rapidly, such as eaves and ceilings of carports and porches. This dew then provides a source of moisture for the mildew.

Mildew fungi can be distinguished from dirt by examination under a high-power magnifying glass. In the growing stage, when the surface is damp or wet, the fungus is characterized by its threadlike growth. In its dormant stage, when the surface is dry, it has numerous egg-shaped spores; by contrast, granular particles of dirt appear irregular in size and shape. A simple test for the presence of mildew on wood and paint can be made by applying a drop or two of liquid household bleach solution (5% sodium hypochlorite) to the stain. The dark color of mildew will usually bleach out in 1 or 2 minutes. A surface stain that does not bleach is probably dirt. It is important to use fresh bleach solution since it deteriorates upon standing and loses its potency.

Before repainting, the mildew must be killed or it will grow through the new paint coat. To kill mildew on wood or on painted wood (or any painted surface), and to clean an area for general appearance or for repainting, use a bristle brush or sponge to scrub the painted surface with the following solution: ⅓ cup household detergent, 1 quart (5%) sodium hypochlorite (liquid household bleach), 3 quarts warm water. When the surface is clean, rinse it thoroughly with fresh water. Allow it to dry before finishing.

Warning: Do not mix bleach with ammonia or with any detergents or cleansers containing ammonia. Mixed together the two are a lethal combination, similar to mustard gas. In several instances people have died from breathing the fumes from such a mixture. Many household cleaners contain ammonia, so be extremely careful in selecting the type of cleaner you mix with bleach.

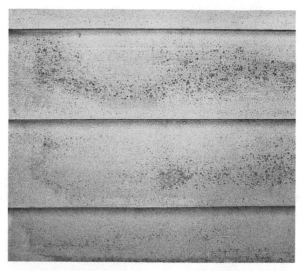

Figure 7-13. Mildew on paint is most common in warm humid climates as well as in shaded or protected areas.

Figure 7-14. A streaked type of water-soluble extractive discoloration can result from water wetting the back of one piece of siding and then running down on the front of the next piece.

7.5.7 Extractives in Wood

In some species of wood the heartwood contains water-soluble extractives, whereas sapwood does not. These extractives can occur in both hardwoods and softwoods. Western redcedar and redwood are two common softwood species used in construction that contain large quantities of extractives. The extractives (Fig. 7-14) give these species their attractive color, good stability, and natural decay resistance, but they can also discolor paint. Woods such as Douglas-fir and southern pine can also cause occasional extractive staining problems. When extractives discolor paint, moisture is usually the cause.

Discoloration of paint occurs when the extractives are dissolved and leached from the wood by water. When the solution of extractives reaches the painted surface, the water evaporates, leaving the extractives as a yellow to reddish-brown stain. The water that gets behind the paint and causes moisture blisters also causes migration of extractives. The discoloration produced when water wets siding from the back frequently forms a rundown or streaked pattern.

If discoloration is to be stopped, moisture problems must be eliminated. The remaining rundown discoloration will usually weather away in a few months. However, discoloration in protected areas can become darker and more difficult to remove with time. In these cases, wash the discolored areas with a mild detergent soon after the problem develops. Paint cleaners are sometimes effective on darker stains.

7.5.8 Iron Stains

Rust (Fig. 7-15) may be one type of staining problem associated with iron. When standard ferrous nails are used on exterior siding and then painted, a red-brown discoloration may occur through the paint in the immediate vicinity of the nailhead.

To prevent rust stains, use corrosion-resistant nails. These include high-quality galvanized, stainless steel, and aluminum nails. Poor-quality galvanized nails can corrode easily and, like ferrous nails, can cause unsightly staining of the wood and paint. The galvanizing on the heads of the nails should not "chip loose" as they are driven into the wood. If rust is a serious problem on a painted surface, the nails should be countersunk, caulked, and the area should be spot-primed and then topcoated.

Unsightly rust stains may also occur when standard ferrous nails are used in association with any of the other finishing systems such as solid-color or opaque stains, semitransparent penetrating stains, and water-repellent pre-

Figure 7-15. Metal fasteners or window screens can corrode and later discolor paint as leaching occurs.

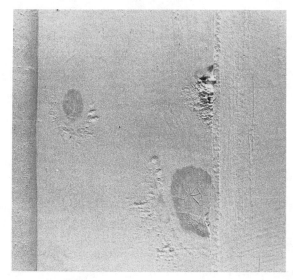

Figure 7-16. Brown discoloration of paint due to resin exudation from a knot.

servatives. Rust stains can also result from screens and other metal objects or fasteners that are subject to corrosion and leaching.

A chemical reaction with iron resulting in an unsightly blue-black discoloration of wood can also occur. In this case, the iron reacts with certain wood extractives to form the discoloration. Woods most prone to this type of discoloration are western redcedar, redwood, Douglas-fir, and the oaks. Ferrous nails are the most common source of iron for chemical staining, but problems have also been associated with traces of iron left from cleaning the wood surface with steel wool or wire brushes. The discoloration can sometimes become sealed beneath a new finishing system.

Oxalic acid solution will usually remove the blue-black chemical discoloration from iron provided it is not already sealed beneath a finishing system. The stained surface should be given several applications of the solution containing at least 1 lb of oxalic acid per gal of water, preferably hot. After the stains disappear, the surface should be thoroughly washed with warm fresh water to remove the oxalic acid and any traces of the chemical causing the stain. If all sources of iron are not removed or protected from corrosion, the staining problem may reoccur.

Caution: Extreme care should be exercised when using oxalic acid since this chemical is very toxic.

7.5.9 Brown Stain Over Knots

The knots (Fig. 7-16) in many softwood species, particularly pine, contain an abundance of resin. This resin can sometimes cause paint to peel or turn brown. In most cases, this resin is "set" or hardened by the high temperatures used in kiln-dried construction lumber.

Good painting practices should eliminate or control brown stain over knots. Apply a good primer to the bare woods first. Then follow with two · topcoats. Special stain-blocking primers are available for use over the knot. Some specially formulated exterior-grade shellacs or pigmented shellacs are available for this purpose. Other types of specialty primers are also available.

7.6 APPLICATION OF WOOD FINISHES

Many of the problems with finishes can be eliminated or minimized by following good coating application practices. Proper surface preparation before applying finishes makes finishes effective. Selecting the right finish for a given substrate is another important factor in ensuring good performance of finished wood surfaces.

7.6.1   Paint

Proper surface care and preparation before applying paint to wood is essential for good performance. Wood and wood-based products should be protected from the weather and wetting during storage on the construction site and after they are installed. Surface contamination from dirt, oil, and other foreign substances must be eliminated. It is most important to paint wood surfaces within one to two weeks, weather permitting, after installation.

To achieve maximum paint life on new wood, follow these steps:

- Wood (siding, trim, fencing, etc.) should be treated with a paintable water-repellent preservative or water repellent, which can be applied by brushing or dipping. Lap and butt joints and the edges of panel products such as plywood, hardboard, and particleboard should be treated carefully because paint normally fails in these areas first. Allow at least two warm, sunny days for adequate drying before painting the treated surface. If the wood has been dip-treated, allow at least one week of favorable weather.
- After the water-repellent preservative or water repellent has dried, the bare wood must be primed. As the primer coat forms a base for all succeeding paint coats, it is very important because it seals in the extractives so that they will not bleed through the topcoat. A primer should be used whether the topcoat is an oil-based or latex-based paint. For woods with water-soluble extractives such as redwood and cedar, the best primers are good-quality oil-based and alkyd-based paints or stain-blocking acrylic latex-based paints.
- Two coats of a good-quality acrylic latex house paint should be applied over the primer. Other paints that are used include the oil-based, alkyd-based, acrylic terpolymer and vinyl-acrylic. Best performance will usually be found with the top-quality all-acrylic latex house paints. If it is not practical to apply two topcoats to the entire house, consider two topcoats for fully exposed areas on the south and west sides as a minimum for good protection. Areas fully exposed to sun and rain are the first to deteriorate and therefore should receive two coats. Window sills will benefit from three coats because of particularly severe exposure.
- One gallon of paint will cover about 400 ft^2 of smooth wood surface area. However, coverage can vary with different paints, surface characteristics, and application procedures. Research has indicated that the optimum thickness for the total dry paint coat (primer and two topcoats) is 4–5 mils or about the thickness of a sheet of newspaper. The quality of paint is usually, but not always, related to price. Brush application is usually superior to roller or spray application, especially for the first coat.

Oil-based paint may be applied when the temperature is 40°F or above. A minimum of 50°F is required for applying latex-based waterborne paints.

For proper curing of these latex paint films, the temperature should not drop below 50°F for at least 24 hours after paint application. Low temperatures will result in poor coalescence (fusing) of the paint film and early paint failure.

To avoid wrinkling, fading, loss of gloss of oil-based paints, or streaking of latex paints, the paint should not be applied in the evenings of cool spring and fall days when heavy dews form during the night before the surface of the paint has thoroughly dried. Serious water absorption problems and major finish failure can also occur with some latex paints when applied under these conditions.

7.6.2 Solid-Color Stains

Solid-color stains may be applied to a smooth surface by brush, spray, or roller application, but brush application is usually best. These stains act much like paint. One coat of solid-color stain is considered adequate for siding by the manufacturer, but two coats will provide significantly better protection and longer service. These stains are not generally recommended for horizontal wood surfaces such as decks and window sills.

7.6.3 Semitransparent Penetrating Stains

Semitransparent oil-based penetrating stains may be brushed, sprayed, or rolled on. Brushing will give the best penetration and performance. These stains are generally thin and runny, so application can be messy. Lap marks may form if stains are improperly applied. They can be prevented by staining only a small number of boards or one panel at a time. This method prevents the front edge of the stained area from drying out before a logical stopping place is reached. Working in the shade is desirable because the drying rate is slower. One gallon will usually cover 200–300 ft² of smooth wood surface and 100–150 ft² of rough or weathered surface.

7.6.4 Water Repellents and Water-Repellent Preservatives

The most effective method of applying a water repellent or water-repellent preservative is to dip the entire board into the solution. However, brush treatment is also effective. When wood is treated in place, liberal amounts of the solution should be applied to all lap and butt joints, edges and ends of boards, and edges of panels where end grain occurs. Other areas especially vulnerable to moisture, such as the bottoms of doors and window frames, should not be overlooked. One gallon will cover about 250 ft^2 of smooth surface or 150 ft^2 of rough surface. The life expectancy is only 1 to 2 years as a natural finish, depending on the wood and exposure. Treatments on rough surfaces are generally longer-lived than those on smooth surfaces. Repeated brush treatment to the point of refusal will enhance durability and performance. Treated wood that is painted will not need retreating unless the protective paint layer weathers away.

7.6.5   Finishing Treated Wood

Wood that has been pressure-treated for decay or fire resistance sometimes has special finishing requirements. All the common pressure preservative treatments (creosote, pentachlorophenol, water-repellent preservatives, and waterborne) will not significantly change the weathering characteristics of wood. Certain treatments such as waterborne treatments containing chromium reduce the degrading effects of weathering. Except for esthetic or visual reasons, there is generally no need to apply a finish to most preservative-treated wood. If needed, water repellents or oil-based semitransparent penetrating stains can be used. These finishes can be applied to clean, dry surfaces of CCA-treated wood, but can be applied to preservative oils or solutions (creosote, pentachlorophenol) only after the wood has weathered for 1 to 2 years depending on exposure. The only preservative-treated wood that can be painted or stained immediately after treatment and without further exposure is CCA-treated wood, but only when the treated wood is clean and dry. Most CCA-treated wood is usually prepared from species with fair to poor paint-holding characteristics. Thus, the most desirable finishes for CCA-treated wood are the oil-based semitransparent stains. Manufacturers generally have specific recommendations for good painting and finishing practices for fire-retardant and preservative-treated woods.

Wood pressure-treated with waterborne chemicals, such as copper, chromium, and arsenic salts (CCA-treated wood), that react with the wood or form an insoluble residue presents no major problem in painting if the wood is properly redried and thoroughly cleaned after treating. Wood treated with preservative oils (creosote, pentachlorophenol solutions) normally cannot be painted successfully especially when heavy oil solvents with low volatility are used to treat wood under pressure.

7.6.6   Finishing Porches and Decks

Exposed flooring on porches and decks is sometimes painted even though painting is not recommended for fully exposed wood. The recommended procedure of treating with water-repellent preservative and primer is the same as for wood siding. After the primer, an undercoat (first topcoat) and matching second topcoat of porch and deck enamel should be applied. Many fully exposed decks are more effectively finished with only a water-repellent preservative or a penetrating-type oil-based semitransparent pigmented stain. These finishes will need more frequent refinishing than painted surfaces, but it is easily done because there is no need for laborious surface preparation as when painted surfaces start to peel. Solid-color stains should not be used on any horizontal surface such as decks because early failure may occur.

7.7 REFINISHING WOOD

Exterior wood surfaces need be refinished only when the old finish has worn thin and no longer protects the wood. Too frequently, application of paint causes the paint film to become too thick, making the finish susceptible to cracking.

7.7.1 Paint and Solid-Color Stains

In refinishing an old paint coat (or solid-color stain), proper surface preparation is essential if the new coat is to give the expected performance. First, scrape away all loose paint. Use sandpaper on any remaining paint to "feather" the edges smooth with the bare wood. Then scrub the exposed wood surface and any remaining old paint with a brush or sponge and water. Rinse the scrubbed surface with clean water. Wipe the surface with your hand. If the surface is still dirty or chalky, scrub it again using a detergent. Mildew should be removed with a dilute household bleach solution. Rinse the cleaned surface thoroughly with fresh water and allow it to dry before repainting. Areas of exposed (bare) wood should be treated with a water-repellent preservative or water repellent and allowed to dry for at least 2 days and then primed. Topcoats can then be applied.

It is particularly important to clean areas protected from sun and rain such as porches, soffits, and side walls protected by overhangs before refinishing, especially repainting. These areas tend to collect dirt and water-soluble materials that interfere with adhesion of the new paint. For semitransparent stains, old stain or other contamination may interfere with penetration. It is probably adequate to refinish these protected areas every other time (or every third time) the house is refinished.

Latex paint can be applied over freshly primed surfaces and on weathered paint surfaces if the old paint is clean and sound. Where old sound paint surfaces are to be repainted with latex paint, a simple test should be conducted first. After cleaning the surface, repaint a small, inconspicuous area with latex paint, and allow it to dry at least overnight. Then, to test for adhesion, firmly press one end of an adhesive bandage onto the painted surface. Jerk it off with a snapping action. If the tape is free of paint, the latex paint is well bonded and the old surface does not need priming or additional cleaning. If the new latex paint adheres to the tape, the old surface is too chalky and needs more cleaning or the use of an oil-based primer. If both the latex paint and the old paint coat adhere to the tape, the old paint is not well bonded to the wood and must be removed before repainting.

7.7.2   Semitransparent Penetrating Stains

Semitransparent oil-based penetrating stains are relatively easy to refinish. Heavy scraping and sanding are generally not required. Simply use a stiff-bristle brush to remove all surface dirt, dust, and loose wood fibers, and then apply a new coat of stain. The second coat of penetrating stain often lasts longer than the first since it penetrates into small surface checks that open up as wood weathers.

7.7.3   Water-Repellent Preservatives

Water-repellent preservatives used for natural finishes can be renewed by a simple cleaning of the old surface with a bristle brush and an application of a new coat of finish. To determine if a water-repellent preservative has lost its effectiveness, splash a small quantity of water against the wood surface. If the water beads up and runs off the surface, the treatment is still effective. If the water soaks in, the wood probably should be refinished. Refinishing is also required when the wood surface shows signs of graying. Gray discoloration can be removed by using liquid household bleach (see section on mildew).

7.8   SUMMARY

A wood finish should protect the surface, help maintain the appearance, and provide cleanability. The type, quality, quantity, and application method of the finish all impact the effectiveness. The type of wood and its condition also influence the success and longevity of the finish. Making informed decisions regarding wood finishes, carefully applying the finish, and maintaining the finish lead to the proper protection and best service of the wood surface.

7.9   SUGGESTED READINGS

Banov, Abel. 1978. *Paints & Coatings Handbook*, 2nd ed. Farmington, MI: Structures Publishing Company.

Banov, Abel, and Marie-Jeanne Lytle. 1975. *Book of Successful Painting*. Farmington, MI: Structures Publishing Company.

Browne, F. L. 1962. *Wood Properties and Paint Durability.* Miscellaneous Publication 629. Madison, WI: U.S. Department of Agriculture, Forest Service, Forest Products Laboratory.

Brushwell, William. 1973. *Goodheart-Willcox's Painting and Decorating Encyclopedia.* South Holland, IL: The Goodheart-Willcox Company, Incorporated.

Cassens, D. L., and W. C. Feist. 1986. *Finishing Wood Exteriors: Selection, Application, and Maintenance.* Agriculture Handbook No. 647. Madison, WI: U.S. Department of Agriculture, Forest Service, Forest Products Laboratory.

Charnow, Will. 1987. *Painter's Pal.* Santa Barbara, CA: The Painter's Pal Company.

Feist, W. C. 1982. Weathering of wood in structural uses. In *Structural Use of Wood in Adverse Environments*, R. W. Meyer and R. M. Kellogg (eds.), pp. 156–178. New York: Van Nostrand Reinhold.

Feist, W. C., and E. A. Mraz. 1978. *Wood Finishing: Water Repellents and Water-Repellent Preservatives.* Research Note FPL-0124. Madison, WI: U.S. Department of Agriculture, Forest Service, Forest Products Laboratory.

Feist, W. C., and A. E. Oviatt. 1983. *Wood Siding—Installing, Finishing, Maintaining.* Home and Garden Bulletin 203. Washington DC: U.S. Department of Agriculture.

Hamburg, H. R., and W. M. Morgans. 1979. *Hess's Paint Film Defects*, 3rd ed. Bungay, Suffolk: Richard Clay (The Chaucer Press), Ltd.

Soderberg, George. 1969. *Finishing Technology*, 3rd ed. Bloomington, IL: McKnight & McKnight Publishing Company.

Weismantel, Guy E. 1981. *Paint Handbook.* New York: McGraw-Hill Book Company.

7.10 ASSOCIATIONS WITH INFORMATION ON WOOD FINISHING

American Hardboard Association
887-B Wilmette Road
Palatine, IL 60067

American Plywood Association
7011 South 19th
P.O. Box 11700
Tacoma, WA 98411

California Redwood Association
591 Redwood Highway, Suite 3100
Mill Valley, CA 94941

Forest Products Laboratory
USDA Forest Service
One Gifford Pinchot Drive
Madison, WI 53705

Forest Products Research Society
2801 Marshall Court
Madison, WI 53705

Hardwood Research Council
P.O. Box 131
Asheville, NC 28802

Maple Flooring Manufacturers Association
Suite 720, South
Sperry Univac Plaza
8600 West Bryn Mawr Avenue
Chicago, IL 60631

National Forest Products Association
1250 Connecticut Avenue, NW
Washington, DC 20036

National Oak Flooring Manufacturers Association
810 Sterick Building
Memphis, TN 38103

National Wood Window and Door Association
1400 East Touhy Avenue
Suite G-54
Des Plaines, IL 60018

Red Cedar Shingle & Handsplit Shake Bureau
Suite 275
515 116th Avenue, NE
Bellevue, WA 98004

Southern Forest Products Association
P.O. Box 52468
New Orleans, LA 70152

Western Wood Products Association
1500 Yeon Building
Portland, OR 97204

8. Exterior Masonry Surfaces

Albert W. Isberner and
Lynn R. Lauersdorf

8.1 INTRODUCTION

Preventive maintenance (PM) of exterior surfaces is a thriving industry. One needs to look only at the number of manufacturers and contractors that specialize in this area to realize its significance. Numerous building failures can be attributed to poorly performing masonry systems. They are second in intensity only to the roofing systems in failure frequency and litigation.

Exterior masonry systems serve as the weather-resisting enclosures of building interiors; in addition, they can be designed to be load-bearing structural elements. Masonry wall systems are predominantly made up of clay units (brick) or concrete units (block) joined with mortar.

Masonry systems are subject to environmental influences, mainly to moisture and temperature fluctuations. Both clay and concrete products experience shrinkage and expansion as the moisture level and temperature of the surroundings fluctuate. All clay products expand irreversibly with the absorption of moisture vapor. Additionally, concrete products shrink upon their reaction with carbon dioxide in the atmosphere. Clay and concrete masonry systems are susceptible to water penetration especially at the mortar joints. These two factors—movement of masonry caused by temperature changes and its susceptibility to water entry—should serve as a basis for the design and maintenance of the masonry systems.

Many of the potential problems with masonry walls can be alleviated through good design, appropriate materials, and good construction practices (Fig. 8-1). Design and detailing of the masonry is very important for obtaining weather-resistant masonry building. The materials selected as well as the quality of workmanship significantly affect the performance and longevity of masonry structure.

The prevention of premature masonry failures, however, cannot be satisfactorily accomplished without PM of the building masonry system. Preventive maintenance of masonry systems as of any other building systems should

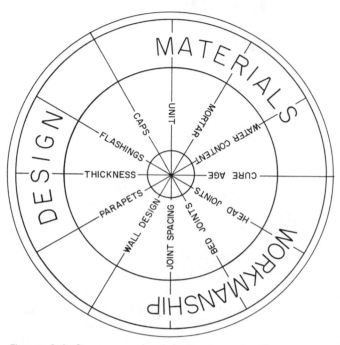

Figure 8-1. Components of design/construction process for ensuring weather-resistant masonry.

start during the project design phase. Both the designer and the owner of a new building must consider PM as one of the primary responsibilities of ownership. While masonry has a well-earned reputation for longevity, it can only be realized if planning for PM of masonry systems starts early in the project design and continues throughout the life of the structure.

8.2   MASONRY WALL TYPES

There are two basic masonry wall types: solid or barrier walls, and cavity walls. Barrier walls consist of masonry units and mortar used to bed the units and fill the collar joint-air space between adjacent wythes (Fig. 8-2). When properly designed and built, these walls provide an effective barrier against moisture.

Cavity walls are constructed of two or more wythes, separated by a cavity (Fig. 8-3). Resistance to water penetration is accomplished by the space between the two wythes. In theory, any water penetrating the exterior wythe of this type of wall enters the cavity and flows downward on the inner surface of the exterior wythe until it is intercepted by flashings that channel the water

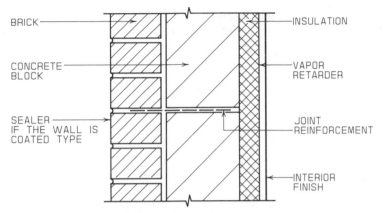

Figure 8-2. Masonry barrier wall.

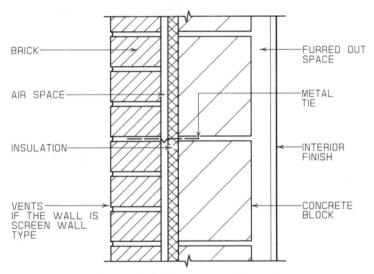

Figure 8-3. Masonry cavity wall.

to the outside through the weepholes. Brick or block veneer walls fall into the cavity wall category. The backing of these walls consists of wood or metal framing.

The success of cavity walls is largely dependent on workmanship. It is critical that mortar droppings are not allowed to build up within the cavity, damming it and blocking the weepholes. If the precautions to keep the weepholes open are not taken, the cavity wall can become totally ineffective.

A further subdivision of these two basic types of masonry walls are screen and skin walls. A screen wall equalizes the pressure between the outside environment and the cavity within the wall system, preventing water entry into the building. The backup wythe serves as the primary air seal, reducing the potential for air leakage and related condensation problems, especially in buildings humidified in the winter. Screen walls are generally the most positive means of preventing water penetration.

A skin wall depends on a seal at the exterior surface to keep water from penetrating into the wall. The sealed wall concept is probably the least satisfactory in prevention of moisture penetration into the masonry wall system. For this system to work, a perfect barrier against moisture is required, which is almost impossible to achieve.

8.3   COMPONENTS OF MASONRY WALL SYSTEMS

Building systems generally do not fail as systems. Failures commence with one or more of the system's components. Maintenance of building systems also does not generally deal with a complete system, but with some of the components of the system. To be able to understand how premature failures of masonry systems can be prevented, it is important to understand the function of the components of the system. Masonry systems are made up of masonry units, mortar, reinforcement, shelf angles, ties, flashings, expansion and control joints, caulks, sealants, and coatings.

The selection of masonry units usually depends on their moisture absorption susceptibility, strength, and the desired visual effect. Masonry wall failures generally do not start with masonry units, unless they are highly permeable to moisture and are exposed to freeze and thaw cycles. Mortar is much more susceptible to premature failures (Fig. 8-4).

Mortar serves many functions. In addition to providing a desired aesthetic effect, the mortar must bond and seal masonry units at joints. During construction, mortar must possess workability, strength, and capability to bond masonry units together. For the ease of workmanship, it should possess adequate plasticity and an appropriate hardening rate.

Partial or localized loss of bond between units and mortar is a relatively common occurrence. This defect appears as a slight opening similar to a crack between the masonry unit and the mortar (top of mortar and bottom of unit

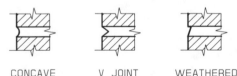

CONCAVE   V JOINT   WEATHERED

Figure 8-4. Recommended masonry joints.

especially). The defect can be caused by the hard, low-suction masonry units too commonly used today and high-cement-content mortars or an incompatible unit and mortar combination.

For brick (or concrete block) and mortar to be compatible, three requirements have to be satisfied. First, units should possess a degree of absorptivity to prevent sliding when they are being placed. This requirement should be ensured especially in cold weather. Second, the mortar should contain enough water to wet the entire masonry unit's bedded surface. If this condition is not met, some unit areas may not bond. Third, the mortar and masonry units should be capable of developing adequate bond (sufficient adhesion capability for units to be held together by mortar) at the face of the mortar and the units. The lack of bond between the units and the mortar can be intensified by a typical acid wash of new or existing construction. Acid wash can be more harmful than long-term weathering.

Masonry systems commonly contain horizontal as well as vertical reinforcement. Reinforcement should be adequately protected by a cover of mortar at its exterior surface, or it should be corrosion-resistant. Reinforcement should always be continuous except at the movement joints.

One major cause of masonry wall failures is inappropriately designed, installed, and maintained flashings. Numerous water problems in masonry systems have resulted from the lack of through-the-wall flashing. Problems also occur with concealed flashing, a flashing held back from the exterior surface of the wall rather than brought out beyond the surface and turned down to serve as a drip. If a drip is not installed, steps should be taken to ensure that the flow of water from within the masonry system to the outside is not impaired. Figure 8-5 shows recommended flashing installations.

To accommodate building movement, as well as the expansion and contraction of masonry, expansion, control, construction, and isolation joints should be installed (Fig. 8-6). Proper design, installation, and maintenance of expansion and control joints are critical.

Expansion joints accommodate expansion of the building sections and actually divide the building into parts. They are often confused with control joints. Control joints (sometimes called contraction joints) are formed, sawed, or tooled grooves to regulate the location and amount of cracking. When the walls begin to shrink, cracks form in the grooves (control joints), since they are the planes of weakness. Joints in concrete masonry should provide a bond break to accommodate both short- and long-term shrinkage and the joints in clay masonry should accommodate long-term expansion. Such joints should also provide for reversible movement due to thermal and moisture content fluctuations.

Size and spacing of expansion and control joints is critical if sealants used to close the joints are to perform properly. Working or dynamic joints (joints that open and close) require maximum performance from the sealant since it

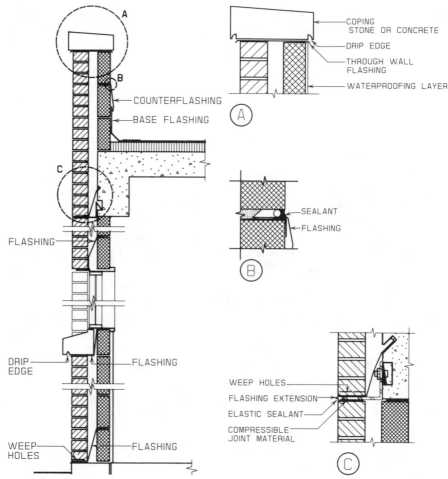

Figure 8-5. Failure-prone areas of masonry wall construction.

must expand and contract with the movement of the joint. Sealants are load-bearing elastic joint materials capable of expanding and contracting with the motion of joints (caulks are not load-bearing). They close an opening in a structure to keep out water, air, and dirt.

Sealants selected for dynamic joints must be able to withstand recurring stresses without failure. The selection of sealants should be based on their ability to function under specified conditions. There are two types of sealant failures: adhesion and cohesion (Fig. 8-7). Adhesion failure indicates a loss of bond between sealants and masonry, perhaps because the masonry was not properly prepared for sealant application. Cohesion failure indicates an

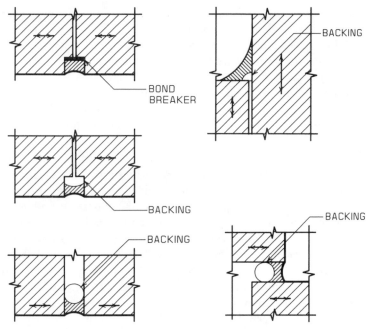

Figure 8-6. Expansion/contraction joints recommended for masonry construction.

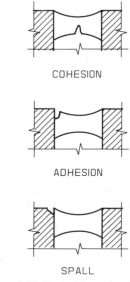

COHESION

ADHESION

SPALL

Figure 8-7. Typical sealant failures.

internal stressing of the sealant, perhaps because of the use of improper sealant or improper sealant configuration.

For caulking or recaulking, proper joint configuration is essential. The cross-section of the sealant should have the depth at the center approximately half its width. Proper joint preparation is also important. All porous surfaces should be primed. Typical closed-cell subcaulking rod stock should be installed under about 25% compression and unscarred after installation. Use bond breaker (polyethylene) tape for alternate situations to avoid a three-sided adhesion.

The extent of movement at joints depends on the size of sections joined, the coefficient of thermal expansion/contraction of the material, and the temperature/moisture fluctuations. The ambient temperature at the time of installation also has some bearing, as does wind load, vibration, and settlement.

Designing the joint properly and choosing and installing the right sealant go hand-in-hand to determine if the joint will work successfully. See Table 8-1 for selecting appropriate caulks and sealants. Generally, experts recommend using polyurethanes for porous materials and for joints between porous and nonporous materials. Silicones are recommended for nonporous materials. Usually, good silicones are not paintable.

Brick masonry is a very durable material and should not be painted or coated unless necessary for some special purpose, such as alleviating graffiti problems. Concrete masonry, on the other hand, does require some surface treatment against moisture penetration and, in many instances, for appearance.

It is critical that the coating does not trap moisture within a masonry wall. Masonry units can fail during freezing and thawing, if the wall is coated with nonporous coatings. A coating that allows moisture to escape generally is not detrimental to masonry surfaces. Coatings should be applied uniformly and with appropriate thickness to ensure complete coverage.

Before applying coatings, it is important to ensure that the surface being coated is clean. Coatings can usually be applied with a brush or spray and normally require two or more layers. Follow the manufacturer's instructions for coating applications. There are numerous coatings available for use on masonry surfaces.

Portland cement paints: These compounds are generally a mixture of portland cement, lime, fine sand, stearate, and an optional coloring compound. These paints develop adhesion and weather resistance through cement hydration, so moist curing promotes and enhances the quality of the paint. The paints are considered excellent toward reducing water penetration and excellent toward vapor transmission.

Latex paints: These compounds include acrylics, polyvinyl acetates, and rubber-based emulsions. Being air-cured systems, they may be applied to damp surfaces and do not require moist curing. They have good vapor transmission characteristics.

Oil-based paints: These compounds combine natural oil resins or synthetic alkyd resins. The oil-based paints are subject to blistering and peeling when the masonry is wet. Wet concrete masonry is very caustic with possible pH values ranging from 12.5 to 14, so discussions with paint suppliers should consider this fact. Some manufacturers increase the resistance to blistering and peeling by altering the composition with resins.

Rubber-based coatings: These compounds include chlorinated rubber, styrene acrylate, vinyl toluene-butadiene, and styrene butadiene.

Epoxy coatings: These coatings contain epoxy resin to which a catalyst is added. An impervious film forms on the surface. Certain formulations have shown a hydrolysis—water intake accompanied with expansional blistering—when water concentrated immediately below the coating surface.

Silane: Relatively new surface treatments for both concrete and masonry involve the application of silanes and siloxanes. These clear, penetrating surface treatments have proven desirable.

When selecting a coating that would allow for an easy removal of such markings as graffiti from masonry surfaces, a number of factors must be examined. Masonry coatings should be capable of forming a film on the surface to which they are applied. They should be able to resist weathering and ultraviolet rays. Additionally, the coatings should be permeable to water vapor but impermeable to water and should be clear so as not to change the appearance of the masonry. The main characteristic of a coating, of course, should be its cleanability.

Of the numerous coatings available on the market, acrylics come closest to satisfying these requirements. Vinyl coatings have adhesion problems and polyurethanes are expensive, which makes these coatings less desirable. Problems with acrylics include their softening characteristics when exposed to water, thus attracting dirt. Acrylics are also susceptible to freezing and could exhibit discoloration when wet.

8.4 CAUSES OF MASONRY FAILURES

Understanding what causes failure of building systems is advantageous toward ensuring that the systems are constructed and maintained properly. Specifying materials and construction procedures without a thorough knowledge of how building systems fail could lead to costly repairs and maintenance in the years to come.

Table 8-1. Characteristics of Sealing Compounds

| | | Butyls | | Acrylics | |
	Oil Base	Skinning Type	Nonskinning Type	Solvent-Release Type	Water Release Typene
Chief ingredients	Selected oils, fillers, plasticizers, binders, pigment	Butyl polymers, inert reinforcing pigments, nonvolatile plasticizers, and polymerizable dryers	Butyl polymers, inert reinforcing pigments, nonvolatizing and nondrying plasticizers	Acrylic polymers with limited amounts of fillers and plasticizers	Acrylic polymers with fillers and plasticizers
Primer required	In certain applications	None	None	None	
Curing process	Solvent release, oxidation	Solvent release, oxidation	No curing: remains permanently tacky	Solvent release	Water Evaporation
Tack-free time (hr)	6	24	Remains indefinitely tacky	36	36
Cure time (days)	Continuing	Continuing	N/A	14	5
Max. cured elongation	15%	40%	N/A	60%	Not available
Recommended max. joint movement	± 3% decreasing with age	± 7½%	N/A	± 10%	± 5%
Maximum joint width	1″	¾″	N/A	¾″	⅝″
Resiliency	Low	Low	Low	Low	Low
Resistance to extension	Very low	Moderate	Low	Very low	Low
Resistance to compression	Very low	Low	Low	Very low	Low
Service temp. range °F	−20° to 150°	−20° to 180°	−20° to 180°	−20° to 180°	−20° to 180°
Normal application temperature range	+40 to +120°	+40° to +120°	+40° to 120°	+40° to 120°	+40° to 120°
Weather resistance	Poor	Fair	Fair	Very good	Not available
Ultraviolet resistance, direct	Poor	Good	Good	Very good	Not available
Cut, tear, abrasion resistance	N/A	N/A	N/A	N/A	N/A
Life expectancy	5–10 years	10 years +	10 years +	20 years +	Not available

Source: From Maslow 1974.

Polysulfides		Polyurethanes		
One-Component	*Two-Component*	*One-Component*	*Two-Component*	*Silicones*
Polysulfide, polymers, activators, pigments, inert fillers, curing agents, and nonvolatizing plasticizers	Base: polysulfide, polymers, activators, pigments, plasticizers, fillers. Activator: accelerators, extenders, activators	Polyurethane prepolymer, filler pigments and plasticizers	Base: polyurethane prepolymer, filler, pigment, plasticizers. Activator: accelerators, extenders, activators	Siloxane polymer, pigment, and fillers
Usually	Usually	Usually	Always	Usually
Chemical reaction with moisture in air and oxidation	Chemical reaction with curing agent	Chemical reaction with moisture in the air	Chemical reaction with curing agent	Chemical reaction with moisture in the air
24	36–48	36	24	2
14–21	7	14	3–5	5
300%	600%	300%	400%	250%
± 25%	± 25%	± 15%	± 25%	± 20%
1″	1″	¾″	1″	⅝″
High	High	High	High	High
Moderate	Moderate	High	High	High
Moderate	Moderate	Moderate	High	High
−40° to 200°	−60° to 200°	−25° to 250°	−40° to 250°	−60° to 250°
+40° to 120°	+40° to 120°	+40° to 120°	+40° to 120°	+20° to +160°
Good	Good	Very good	Very good	Excellent
Good	Poor to good	Poor to good	Poor to good	Excellent
Good	Good	Poor to good	Poor to good	Excellent
20 years +	20 years +	20 years +	20 years +	20 years +

The main thrust of building design and construction, in addition to providing a well-functioning facility for the user, should be a facility built to withstand the elements (Fig. 8-8). Although all buildings have to be maintained to prevent premature failures due to environmental influences, unnecessary and expensive maintenance should not be built into building systems.

Understanding the causes of masonry failures can serve as a guide toward design and construction of a trouble-free masonry system. By knowing the

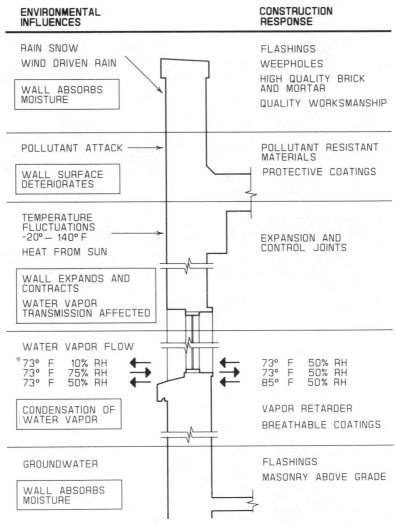

ENVIRONMENTAL INFLUENCES

RAIN SNOW
WIND DRIVEN RAIN

WALL ABSORBS MOISTURE

POLLUTANT ATTACK

WALL SURFACE DETERIORATES

TEMPERATURE FLUCTUATIONS
-20° – 140° F
HEAT FROM SUN

WALL EXPANDS AND CONTRACTS

WATER VAPOR TRANSMISSION AFFECTED

WATER VAPOR FLOW

73° F   10% RH
73° F   75% RH
73° F   50% RH

CONDENSATION OF WATER VAPOR

GROUNDWATER

WALL ABSORBS MOISTURE

CONSTRUCTION RESPONSE

FLASHINGS
WEEPHOLES
HIGH QUALITY BRICK AND MORTAR
QUALITY WORKSMANSHIP

POLLUTANT RESISTANT MATERIALS
PROTECTIVE COATINGS

EXPANSION AND CONTROL JOINTS

73° F   50% RH
73° F   50% RH
85° F   50% RH

VAPOR RETARDER
BREATHABLE COATINGS

FLASHINGS
MASONRY ABOVE GRADE

Figure 8-8. Illustration of masonry design/construction responses to environmental elements.

cause of the potential defect, in many instances, the problem can be corrected before it occurs. Lacking this knowledge could force continuous maintenance of a recurring failure.

8.4.1 Moisture

The major cause of failure of masonry systems is water vapor, liquid, or ice. Water penetration through the masonry wall can cause damage not only to the masonry system itself and to other building systems but also to the contents of building interiors (Fig. 8-9). Water entry through the masonry

Figure 8-9. Damage to interiors caused by water entry through the masonry wall. *(Courtesy: Division of State of Wisconsin Facilities Management.)*

Figure 8-10. Brick joint deterioration and unit spalling. *(Courtesy: Division of State of Wisconsin Facilities Management.)*

Figure 8-11. Spalling of masonry units. *(Courtesy: Division of State of Wisconsin Facilities Management.)*

Figure 8-12. Spalling of masonry units. *(Courtesy: Division of State of Wisconsin Facilities Management.)*

wall can cause corrosion, decay, dimensional change, staining, increased heat transmission, and efflorescence to the components of masonry systems.

Water entry into the masonry system is caused by rain penetration or condensation of moisture on masonry surfaces or within construction. Rain penetration of masonry walls is one of the most frequent complaints of building users. Design using barrier or cavity walls lessens this problem as do good construction practices and good materials.

Condensation could also be responsible for numerous concerns. Condensation in a wall system is generally most noticeable in late winter or very early spring and contributes significantly to efflorescence. It results from the cooling of water vapor that can be carried by air currents or by the vapor pressure differential.

The ultimate cause of failures of masonry systems is freezing and thawing of the accumulated moisture. When the confined water freezes it expands with pressures approaching 15,000 psi. The water expands about 10% as it changes into ice. The higher the moisture content in the system, the greater the likelihood of damage. Severe weathering of mortar joints (Fig. 8-10) and spalling of masonry units (Figs. 8-11 and 8-12) are usually the early indications of such attack.

Because exposure to the elements is more severe at higher building elevations, freeze-thaw damage tends to occur in these areas first. Extremely soft brick (common brick) as well as glazed brick are highly susceptible to freeze-thaw damage known as spalling. Spalling of masonry units under these conditions can be expected, especially in older structures. The masonry at grade or below grade in unheated buildings, in parapet walls and in retaining walls is also highly vulnerable to spalling.

A critical point for water entry and eventual freezing occurs at joints between masonry units and mortar. Water can enter through improperly constructed mortar joints or cracks and can penetrate an opening in the masonry wall of $5/1000$ of an inch.

All single-wythe masonry, regardless of materials or workmanship, has a potential for water penetration. Through an appropriate design this problem can be minimized. A surface treatment may lessen the quantity of water penetration; however, a totally sealed wall is almost impossible to attain. Exercise caution against sealing the wall so as to prevent moisture transmission.

One of the effects of water in the masonry system is efflorescence (Fig. 8-13). Efflorescence is a crystalline deposit, usually white, of water-soluble salts on the surface or in the pores of masonry. Where excessive efflorescence occurs, the force of crystallization can cause disintegration of the masonry. For efflorescence to develop, there must be a source of soluble salts, moisture to act as a vehicle to carry the salts to the surface, and a pressure differential to cause the movement of salt solution. The principal objection to surface efflorescence is the appearance of the salts. Under certain circumstances, particularly when exterior coatings are present, salts can be deposited between

Figure 8-13. Efflorescence: a deposit of water-soluble salts. *(Courtesy: Division of State of Wisconsin Facilities Management.)*

the surface of the masonry units and the coating resulting in unit spalling. Efflorescence is usually found directly below the point of moisture entry that can be caused by rain or condensation. Localized efflorescence is most frequently found at roof to wall junctures or at ground elevations. If the efflorescence is present near ground elevation, the most probable causes are that water is being absorbed from the ground into the masonry or that weepholes are not functioning. Through-the-wall masonry flashings will prevent efflorescence caused by absorbed and rising groundwater. Although selection of masonry construction materials having a minimum of soluble salts is desirable, the prevention of moisture migration through the wall holds the greatest potential in minimizing efflorescence.

8.4.2 Movement

Movement of masonry components is the second largest cause of masonry problems. Virtually all building materials will expand and contract when exposed to daily as well as seasonal temperature changes (Table 8-2). The magnitude of potential thermal movement depends on numerous factors such as air temperature, prevailing wind, and orientation of the building to the sun.

The two most important predictors of thermal movement in exterior masonry are the system's thermal conductance and its color. In certain regions of the Midwest, for example, masonry surface temperatures of $-20°$ F in the winter and $120–160°$ F in the summer are not uncommon.

Building movement, whether caused by shrinkage or expansion of masonry products, building settlement, or its response to thermal changes, can potentially cause numerous masonry failures. These failures manifest themselves

Table 8-2. Causes and Results of Movement of Building Materials

Cause	Result	Clay Masonry	Concrete Masonry	Stone	Concrete	Steel	Wood
Load application	Elastic deformation	yes	yes	yes	yes	yes	yes
Sustained load	Creep contraction	yes	yes	yes	yes	no	yes
Temperature change	Expansion/ contraction	yes	yes	yes	yes	yes	yes
Moisture content change	Expansion/ contraction	yes	yes	yes	yes	no	yes
Cement hydration/ carbonation	Permanent shrinkage	no	yes	no	yes	no	no
Moisture absorption	Permanent expansion	yes	no	no	no	no	no
Freezing	Expansion	yes	yes	yes	yes	no	no

Figure 8-14. These types of vertical and horizontal cracks are often caused by bowing or warping of brick masonry areas. The vertical crack in this illustration is along the control joint. *(Courtesy: Division of State of Wisconsin Facilities Management.)*

Figure 8-15. Vertical crack caused by brick masonry expansion toward viewer. Brick wall rests on concrete foundation without a slip joint, causing the spall in the foundation. *(Courtesy: Division of State of Wisconsin Facilities Management.)*

primarily in terms of cracked masonry surfaces (Fig. 8-14). Movement can also cause spalled foundation corners and shifting of the exterior masonry walls away from the structure (Fig. 8-15).

Expansion joints are used to combat these problems. Walls exposed to the elements on both sides, such as parapets, usually experience greater thermal movement. Thus, more frequent cracking occurs in those locations. Cracks in masonry may occur early after construction or later after loads have been applied and material shrinkage has occurred. Cracks in the masonry systems can contribute significantly to a rapid deterioration of a system by allowing the moisture to penetrate into the system.

The settlement of the structure may not be uniform; the expansion of brick masonry and the shrinkage of concrete masonry, though predictable, may be in excess of the controls incorporated in the design. The long-term deformation/creep of the frame along with moisture and thermal changes may also affect the performance of the exterior masonry facade. Cracks at change of materials are common.

Vertical cracks in the center of walls are usually created by overloads that cause failure by compression. Vertical cracks near the corners are generally created by rotation and displacement of materials in those locations. Horizontal and diagonal cracks on the face of masonry walls are, generally, most intense in the vicinity of the openings in the masonry system. Improperly designed and constructed expansion and control joints fail to meet the demand of movement in masonry.

8.4.3 Design and Construction

As stated earlier, masonry problems resulting from poor design and careless construction practices are numerous. The masonry system's capability to resist penetration of water and to accommodate movement should be prime design considerations. Failure of sealants, extensive loss of bond between units and mortar, misalignment of masonry at control joints or at corners, obstructed weepholes, and inappropriately functioning flashings and expansion joints are attributable to poor design and construction practices.

Appropriately designed masonry systems will reduce the effort needed for PM, will increase the building's longevity, and will reduce its life cycle costs. To ensure good design, consider the following points:

1. Design should discourage unit masonry for copings or sills (Fig. 8-16) and for any projections without durable wash surfaces or flashing coverings.
2. Any masonry system should always contain adequate expansion and control joints with oversize joints in clay products to accommodate swelling of clay masonry.

Figure 8-16. Deteriorating brick masonry sill due to water penetration and freezing in the joints. *(Courtesy: Division of State of Wisconsin Facilities Management.)*

3. A foundation ledge should be provided as a base support for exterior masonry wythe.
4. Masonry should start above-grade and should contain necessary weepholes, flashings, and counterflashings.

Specify lapping and sealing of all flashing joints. Use corrosion-resistant or coated metal. All parapet walls should be flashed through the wall with wall capping pitched to provide adequate drainage. Provide through-the-wall flashing at base supports of the exterior-facing wythe, as well as through-the-wall flashing at all intermediate masonry weight supports such as shelf angles, and at the heads and seals of all wall openings to direct water away from the interior (Fig. 8-5). The exterior edge of such flashing should be adequately exposed and turned down to serve as a drip. The concealed portion of the flashing should be turned up with the interior top edge placed in a reglet or terminated in horizontal joints of the backup masonry wythe. End dams should be provided at the longitudinal ends of flashing over lintels, at column abutments, and adjacent to expansion or control joints. Where flashing occurs at the bed joint, there should be mortar both below and above the flashing, unless the flashing is continuously bonded with adhesive to one of the adjacent materials.

All caps and sills should contain appropriate overhangs with drips and flashings below. Provide continuous metal reinforcement except at expansion joints. Do not use common clay brick for backup or waterproofing coatings that do not allow for migration of moisture.

Planters, decorative brick ledges, and various types of projections of masonry units are often found in masonry construction. They are frequently responsible for numerous masonry failures. Avoid these features especially in potential freeze-thaw areas.

Special emphasis should be placed on design and construction of expansion and control joints (Fig. 8-6). These components of masonry construction are prone to failure and can require extensive PM.

Expansion joints are usually necessary when there are:

1. Adjacent buildings
2. Abutting walls
3. Structural system joints
4. Corners and offsets in walls
5. Junctions of walls in nonrectangular buildings such as L-, T-, or U-shaped
6. Multistory buildings (expansion joints at least on every other level and corners)
7. Framing changes in material or direction
8. Decking changes in material or direction
9. Connections between the building and an addition
10. Intersections of deck and a nonload-bearing wall
11. Deep, long span decks where rotations cause significant motion

To properly design and install expansion joints, analyze existing conditions, select appropriate design details, select appropriate materials, and ensure quality workmanship. The size of the joint is based on the amount of movement expected and the ability of the joint sealant to move.

Whether in PM work or in new construction, workmanship influences not only the performance of the system but the ease or difficulty of its maintenance and associated costs. Careless workmanship during initial construction could prove to be very costly in maintenance and repairs. During construction, masonry materials and walls being erected should be protected from rain and from water used for construction.

It is good practice to employ only specialists in the field. Do not allow for any freezing of masonry systems for at least 24 hours after the construction. Masonry should never be laid during cold weather without appropriate heated materials, wind breaks/enclosures, and heat. It is imperative that complete bedment is provided for masonry units. Masonry should be laid with clean faces. Cleanliness both during and after laying of units reduces the need for acid-washing after construction, which could be damaging to the newly constructed system. Flashings must be properly installed, lapped, and sealed and free of any punctures. Material samples must be approved before work begins.

8.5   PREVENTIVE MAINTENANCE

All buildings and building systems, regardless of the construction quality, need to be maintained regularly to ensure that premature failures do not occur. A PM program should be planned to deal with regularly needed maintenance and should also be responsive to the unpredicted occurrences. Design and construction documents showing the original construction of the building and its modifications should be readily available. This information is invaluable for the maintenance of the building. As-built drawings should also be available.

Important considerations in masonry maintenance include (1) establishing the need for maintenance; (2) identifying the causes of the problem; (3) making a proper diagnosis of the problems; (4) prescribing appropriate maintenance and repair procedures.

8.5.1   Inspection

Inspection of buildings and their components plays a key role in PM of any building systems. Masonry systems should be inspected annually with records of inspections kept for future reference (Fig. 8-17).

Any defects to the masonry system should be precisely located. Compare deterioration noted during earlier inspections with the latest inspection to establish the rate of failure development. Note the location of the areas for further and more detailed inspections. When trying to determine a source of water entry, hose testing may be advisable. If this type of testing is to be performed, start at the base of the building observing for any possible water entry especially in those areas where leaks were reported.

When inspecting masonry systems, look for potential problems resulting from poor construction, design, and inadequate maintenance of masonry systems. Additionally, look for problems caused by moisture, building movement, freeze and thaw cycles, and normal weathering. When inspecting the building for moisture damage, look for efflorescence, leakage, and deteriorating coatings on the masonry surfaces. Water entry or condensation can also cause rusted or rotted frames (Fig. 8-18) in the openings, deterioration of exposed lintels, and failure of inside finishes. Damage caused by freezing and thawing of moisture can show up as surface erosion, cracks, or spalled areas of masonry units and mortar.

Inspection of potential failures of masonry should be approached in a systematic manner. In addition to focusing on the causes of masonry failures, inspection can be organized around major features of masonry construction. These should include parapets, windows, joints, flashings, and wall surfaces. Parapets are a very sensitive portion of the structure because of their exposure to the environmental influences on both sides. Look for problems in parapet capping and for potential failures of parapet wall surfaces as discussed under wall surface degradation. Inspect masonry openings for flashings at the head sill and jam locations. Look for any cracking at head, sill, and corners of masonry openings. When inspecting flashings, make sure flashings are not dammed. Flashings installed without end dams may lead to water damage at the end of the flashing, especially above windows and doors. Flashings may be accompanied by weepholes that are spaced too far apart or are obstructed.

When inspecting movement joints, check to make sure that no adhesion or cohesion failure is evident. Check to ensure that there is an adequate number of expansion joints and that these joints are functioning properly. Working joints exhibit signs of bulging and stretching of sealant. A properly performing sealant adheres to the sides of the joints and is void of cracks.

A number of potential failures can reveal themselves as a result of wall surface inspections. Look for the following potential problem areas when inspecting masonry wall surfaces:

1. Bulging walls in one or more directions
2. Deterioration of brick surfaces by flaking
3. Mortar joint surface deterioration
4. Brick surface cracks and spalls
5. Salt deposits localized or over large areas
6. Coating deterioration at mortar joints and other areas
7. Rust stains every 8, 16, or 24 in. horizontally
8. Rust stains at lintels
9. Mortar spalls from nose of lintel
10. Brick spalls with white soft material at bottom of crater
11. Horizontal crack in masonry wall and vertical cracking

Surface accumulation of leachates (heavy efflorescence) below weepholes and flashing is an indication that masonry above is being penetrated with water, but the weepholes and the flashing are performing satisfactorily. If efflorescence is not present in these areas, penetration of water is not severe or weepholes may be obstructed.

Failures of lintels can occur in a number of areas. Look for rust on the lintel or in the immediate vicinity (Fig. 8-19). If rust is present, inspect the weepholes and flashing immediately above the lintel to make sure they are functioning. If deterioration is taking place, look for cracks at the corners due to lintel deflection and for spalling of mortar joints at the nose of the lintel due to expansion of corroding metal (Fig. 8-20).

Although efflorescence may not be damaging to the structure other than being unsightly, it can point to the areas that contain excessive amounts of moisture that can be damaging to masonry. Location of efflorescence at the ground level indicates that water is being absorbed from the ground. The

EXTERIOR WALLS & OPENINGS REPORT

Building ID No. _____

Agency _____

Building Name _____   Location _____

Inspected by _____   Date _____

Type(s) of Exterior Walls:

☐ Brick ☐ Precast Concrete ☐ Stone ☐ Block ☐ Metal Panels ☐ Wood Siding ☐ Other: _____

INSPECTION ITEM	DEFECTS		DESCRIPTION OF DEFECTS	No Action Will Observe	Corrected by Owner	DSFM Action Requested
	None	Noted				
MASONRY						
Masonry Units Deteriorated						
Mortar Deteriorated						
Extensive Loss of Bond Between Units & Mortar						
Exterior Masonry Moving Away From Structure						
Misalignment Of Masonry At Control Joints or Corners						
Bulging of Masonry Between Control Joints						
Sealant Failure in Control Joints or Other						
Masonry Cracked						
Foundation						
Corners Spalled						
Signs of Water Penetration						
Signs of Efflorescence						

Water Leakage							
Excessive Air Leakage							
Signs of Condensation							
Frames Rotted or Rusted							
Doors/Windows Racked							
Torn/Compressed Calking At Window Jambs or Other							
Exposed Lintels Deteriorated							
Sills Deteriorated or Inadequate							
Paint or Coatings Deteriorated							
Walls Excessively Cold							
Inside Finishes Damaged From "Envelope" Deficiencies							

DOORS-WINDOWS

Figure 8-17. Maintenance checklist for masonry systems. *(Courtesy: Division of State of Wisconsin Facilities Management.)*

EXTERIOR WALLS & OPENINGS REPORT (Continued)

Building ID No. _____

INSPECTION ITEM	DEFECTS None	DEFECTS Noted	DESCRIPTION OF DEFECTS	No Action	Will Observe	Corrected by Owner	DSFM Action Requested
FINISHES - APERTURE OPERATION							
Metal Panels Rusting							
Wood Siding Rotting/Peeling							
Fascia/Soffit/Trim Needs Work							
Door Operation Improper							
Door Hardware Faulty							
Door Weatherstripping Inadequate							
Window Operation Improper							
Window Hardware Faulty							
Window Weatherstripping Inadequate							
Storm Window Problems							
Screen Problems							

OVERALL THE WALLS ARE IN   □ G   □ F   □ P   CONDITION

OVERALL THE DOORS ARE IN   □ G   □ F   □ P   CONDITION

OVERALL THE WINDOWS ARE IN   □ G   □ F   □ P   CONDITION

REMARKS:

240

STRUCTURAL SYSTEM REPORT

Type of Structural System:
□ Reinforced Concrete   □ Steel   □ Wood Frame

INSPECTION ITEM	DEFECTS		DESCRIPTION OF DEFECTS	No Action Will Observe	Corrected by Owner	DSFM Action Requested
	None	Noted				
Columns						
Beams & Framing						
Typical Floor System						
Roof Deck & Support						
Loadbearing Walls						
Foundation Walls						

OVERALL THE STRUCTURAL SYSTEM IS IN   □ G   □ F   □ P   CONDITION.

REMARKS:

Figure 8-17. (Continued)

241

Figure 8-18. Deterioration of frames resulting from a failed caulking between the frame and the masonry surface. *(Courtesy: Division of State of Wisconsin Facilities Management.)*

brick may be too close to the grade or the flashing may have been omitted. If flashing is present but efflorescence appears near the grade, it may mean that the weepholes are obstructed.

Efflorescence occurring near the juncture of the wall and the roof may signal that water enters the wall from the roof. Base flashing, counterflashing, and any other possible roof-related water entry and condensation areas should be carefully investigated. Efflorescence in the vicinity of openings in the masonry may mean that water enters the system through the framing of these openings.

Look for spalling, cracks, and scaling of the units, which means water is penetrating the units. Evidence of cracks at the interface of mortar and brick, cracks within mortar, and scaling or bulging of mortar joint surfaces indicates water penetration into the joints. When inspecting expansion joints, look for adhesion and cohesion failures (Fig. 8-7).

Masonry in wall caps and sills should be checked as should masonry immediately below coping, caps and sills that do not have overhangs, drips, and flashings associated with them, for these areas are highly susceptible to moisture damage.

Figure 8-19. Corroding lintel over opening. *(Courtesy: Division of State of Wisconsin Facilities Management.)*

Figure 8-20. Spalling of mortar in front of steel lintel due to corrosion of the lintel. *(Courtesy: Division of State of Wisconsin Facilities Management.)*

8.5.2   Cleaning

In the life of a masonry structure, it is very likely that it will have to be cleaned. If a building is located in a large city, the cleaning undoubtedly will be performed to remove dirt and a variety of contaminants. Cleaning may also be required if coating of the masonry surface is contemplated. The cleaning method should be based on the type of dirt to be removed and the type of masonry on the building.

The first time a masonry building is cleaned is usually immediately after its completion. Excess mortar and efflorescence are removed at this time. Although, in some cases, chemical cleaning compounds have to be used, their use should be approached with caution.

The simplest masonry cleaning methods are by the use of compressed air, water, water under pressure, water with detergent, or steam. Whenever cleaning is done under pressure, pressure levels must be carefully monitored so masonry surfaces are not damaged.

The decision whether to clean a masonry building mechanically or chemically is controversial. It appears that mechanical cleaning procedures such as sandblasting are not as damaging to the masonry as some experts had assumed. Chemical cleaning, on the other hand, can be destructive if high acid concentrations are used. Also, whether a removal of die skin of units is detrimental has been questioned. High-quality masonry should be able to withstand water and moisture penetration whether the die skin is intact or not. Some experts feel that the need for die skin to resist water has been overrated, especially since it is relatively thin. The protection that it offers to the masonry units is minimal at best.

Generally, light dry or wet sandblasting is desirable rather than washing of masonry with some acid solutions. Certain acids, especially in strong concentrations, leave stains or have difficulty drying out unless special precautions are taken. Sandblasting, on the other hand, pollutes surroundings and has been prohibited in many areas.

Of course, there are other forms of mechanical cleaning. Wire brushing is effective for removing excess mortar or peeling paint. Removal of efflorescence from the facade of masonry also can frequently be achieved by dry brushing. Since many salts are highly soluble in water, they will dissolve under normal weathering process. Where heavy efflorescence accumulates, remove it by using high-pressure water spray or dilute muriatic acid solution. If using muriatic acid, keep its concentration low to avoid damage to mortar and masonry units. Do not use acid on concrete masonry. Light sandblasting, however, is acceptable for removing efflorescence from concrete block.

When using hydrofluoric acid, special procedures should be followed. Although it is an effective cleaning agent for glazed products, its use on unglazed masonry is generally not recommended.

Removal of graffiti can involve a number of procedures depending on what material was used for defacing the masonry surface. Experiment with the simplest method first: water with detergent. Mineral spirits may be effective for some markings. In some cases, stronger solvents such as ketones, lacquer thinners, and paint removers may have to be used. If masonry has been sealed, the worn-out sealant coatings (caused by weathering or by cleaning) can be easily maintained. To locate the areas over which the coating is worn off, check for where the masonry absorbs water.

8.5.3 Mortar Joint Maintenance

In the masonry systems, failures usually start with the mortar joints. There are three maintenance options when mortar joint repairs are indicated. Joints can be tuckpointed, mask grouted (Figure 8-21), or the masonry can be rebuilt.

Mask grouting may be required if substantial amounts of joints are deteriorated. The decision depends largely on economics. If deterioration of brick joints is approaching 50%, mask grouting should be used. This method involves masking of the face of each masonry unit and grouting the joints.

Tuckpointing consists of replacing deteriorated joints individually without affecting the face of the unit. It requires the removal of the old mortar to the depth of approximately ¾ in. New mortar is then placed into the joint. The mortar used for tuckpointing should be allowed some setting after mixing and

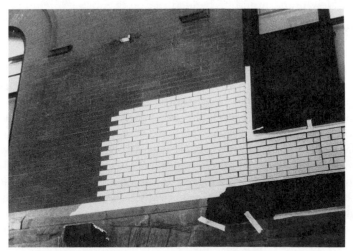

Figure 8-21. Grouting of deteriorated joints. *(Courtesy: Division of State of Wisconsin Facilities Management.)*

before being used. This delay is to minimize shrinkage of mortar after its installation into a joint.

Conditions of mortar joints can be determined by a scratch hardness test. By using a sharp instrument such as a nail, and by scratching mortar in the protected area and then in the problem area and comparing the two, the degree of mortar deterioration in the problem area can be established. Additionally, deteriorating mortar expands outward or upward, dislodging overlaying courses. Slight surface dusting of old mortar is common and not objectionable, provided such weathering does not penetrate too deeply.

Rebuilding of masonry is seldom necessary if units are not damaged.

8.5.4 Repairs

Preventive maintenance activities of masonry systems include a variety of minor repairs. Although major repairs of building systems generally do not fall under PM, numerous lesser repairs belong in the PM category.

If a masonry building experiences foundation settlement or if it lacks properly constructed expansion joints, cracking is likely to occur. If cracks are not repaired without delay, they will tend to get wider. Water entering through cracks can cause problems requiring major repairs. When closing cracks that extend through masonry units, it is necessary to ensure that the loading capability of a unit is restored. Epoxy injection is a good technique for solving this problem.

Cracked units under load-bearing members such as lintels should be replaced. New units should be bedded in epoxy mortar to eliminate recurrence of cracking. If epoxy mortar is recessed to a sufficient depth, conventional mortar can be installed outwardly from the epoxy mortar to match existing color and texture. Masonry systems can fail due to spalling. Spalled brick is usually not repairable. Units should be replaced or the spalled wall area should be covered with another material. Masonry sections with stepped cracks suggest inappropriate expansion joints. Saw new expansion joints at the corners, one header or stretcher unit away from the corner. These new expansion joints should be sized properly, backed with backing rod, and sealed.

Movement of two or more adjacent materials with different coefficients of expansion can cause significant damage to a component connected to the two materials unless provisions are made to accommodate the movement differential. Torn or compressed caulking at doors, windows, and other openings is a good example. Different amounts of movement between masonry walls and foundations can cause racked doors and windows or spalled foundation corners if no bond break is provided over the foundation.

As soon as problems resulting from the above deficiencies become likely, the conditions have to be corrected. Joints to accommodate the movement differential between different materials should be installed. If sealant failure is indicated (Fig. 8-22), replace the sealant. The most common sealant failures result from:

1. Improper mixing of components
2. Improper cleaning of joint faces
3. Improper placement of backing rod (the material behind the sealant used to control depth or width-to-depth ratio)
4. Lack of primer
5. Improper tooling
6. Installation of sealant at excessively high or low temperatures

When repairing an expansion joint, all sealant showing adhesion or cohesion failures should be removed. Various grinders and vibratory tools that slice the sealant at the interface are available to ease sealant removal. Before sealant replacement, the surfaces of the joint should be cleansed to remove all traces of the former sealant. After sealant removal, the joint should be examined to ensure that the proper joint has been installed. A control joint

Figure 8-22. Sealant failure. Spalling of the unit at the corner of the opening is due to water entry through the joint. *(Courtesy: Division of State of Wisconsin Facilities Management.)*

may have raked mortar behind the sealant. An expansion joint should be functional throughout the entire thickness of the wall.

Do not trap water by sealing weepholes. This rule becomes extremely important at shelf angle supports. Where such conditions are found, any sealant acting as a dam should be removed to prevent corrosion of shelf angles as well as corrosion of their connections to the structural frame.

If the weepholes are not working, and if rodding is inadequate for opening them, install new weepholes by drilling holes in the masonry. Exercise caution not to puncture the flashing. If flashings are not installed at the foundation and masonry juncture, the best remedy is to install additional weepholes in the masonry just above ground level, which can be done by drilling. The weepholes will reduce the water content of the masonry wall and lessen the efflorescence.

Lintels and shelf angles that are rusting may rotate, causing fracture of supporting masonry units and spalling of the mortar. Remove rust by sandblasting corroded lintels. Ensure that surrounding areas are protected when sandblasting. After removing the rust, recoat the lintel, replace mortar (if mortar was used) with caulk or sealant, and provide appropriate weepholes.

Another often overlooked construction detail is the provision for adequate lateral support of masonry walls. Inadequate lateral support of masonry, accompanied by masonry volume increase and movement of the structure, particularly of roof slabs and exposed columns, causes masonry panels to move out. This condition occurs mostly in the masonry panels that do not have proper ties to the adjacent structural columns. This defect appears as cracks at the juncture of the exterior wall and interior cross walls. It may also appear at the shelf angles over the openings in the masonry walls. Providing positive lateral support by mechanical means has often solved this problem and, in some instances, brought the masonry panels back to their original position.

When buildings are renovated and insulation is added to the interior of outside walls (cold walls hold more moisture than heated walls), a serious problem can occur in the masonry wall system if this procedure is not done with extreme caution. Often, many older buildings with solid masonry walls cannot tolerate insulation being placed on the inside of the exterior walls, especially when freezing and thawing cycles occur. When insulation is added in these situations, be sure that no moisture is trapped in the masonry system. Insulation should be installed during dry and warm times of the year, and adequate vapor barriers should be provided so moisture does not migrate from the building interior toward the exterior and condense within the wall.

8.6   SUMMARY

Masonry construction has been in existence for centuries, yet its failure rate is second only to built-up roofs. One reason for this high rate of degradation is that the masonry construction is complex, consisting of a number of components: masonry units, mortar, flashings, reinforcement, and numerous types of joints. Much of the material used for these components is readily susceptible to environmental attack.

Two main causes of masonry failure are water, especially if the masonry is susceptible to freezing weather, and temperature fluctuations that cause movement. When masonry systems are designed and constructed, moisture penetration into masonry and movement due to temperature fluctuations should always be kept in mind. Masonry walls are never waterproof. Water, especially wind-driven rain, will always penetrate into the system. Proper protection against water entry and movement should always be provided or failures will appear soon after construction.

Appropriate materials and quality workmanship also play an important role in masonry construction, but as important as these factors are, masonry must also be maintained. Preventive maintenance staff must know how to inspect, evaluate, and choose appropriate PM and repair approaches.

The first signs of failure of masonry systems will probably appear in the mortar joints. Trapped moisture can cause spalling of brick and corrosion of metal as on lintels and reinforcement. Expansion and control joints are also prone to failure as are masonry units themselves if water is trapped in them and freezing occurs. Regular inspection for potential failure and maintenance of masonry systems can prevent serious damage.

8.7 SUGGESTED READINGS

Technical Notes on Brick Construction. Brick Institute of America.

Cleaning Brick Masonry	nos. 20, 14, 23
Colorless Coatings	nos. 7E, 14B
Differential Movements	nos. 14A, Rev. II, 18 Series
Efflorescence	nos. 20, 23 Rev., 23A
Flashings	no. 7A Rev.
Floors	nos. 14 Rev., 14B
Foundation Walls	no. 7 Rev.
Maintenance	nos. 20 Rev., 7F, 14 Rev., 14B
Moisture Control	nos. 6, 8, 6B
Painting	no. 6 Rev.
Parapets	no. 18
Stains	no. 20 Rev.

TEK Notes, an Information Series. National Concrete Masonry Association.

Cleaning	nos. 17A, 28, 45, 92, 100
Coatings	nos. 10A, 55, 100, 1212, 149
Condensation	nos. 1, 13A
Corrosion Protection	no. 136
Crack Control	nos. 3, 44, 53
Efflorescence	nos. 44, 92
Flashing	nos. 13A, 62, 126
Graffiti Removal	no. 100

Maintenance   nos. 29, 44, 92, 100
Moisture   nos. 1, 4B, 20A, 40, 54, 59, 60, 107, 143
Paints and Painting   nos. 10A, 18A, 39, 44, 55, 100, 7A
Painting   no. 7A
Sealants   nos. 13A, 126
Stains & Removal of   nos. 1, 4B, 8, 10A, 13, 55, 121, 126
  Waterproofing
Water Tightness   nos. 43, 44, 62, 77, 85, 126
Weepholes   nos. 2A, 67, 72, 52, 141

Gans, F. O. 1977. *Guide to Joint Sealants for Concrete Structures.* ACI 504R-77, Report by ACI Committee 504.

Maslow, Philip. 1974. *Chemical Materials for Construction.* Farmington, MI: Structures Publishing Co.

Panek, J. R., and J. P. Cook. 1984. *Construction Sealants and Adhesives,* 2nd ed. New York: John Wiley and Sons.

8.8   ASSOCIATIONS WITH INFORMATION ON MASONRY CONSTRUCTION

American Concrete Institute
Box 19150 Redford Station
Detroit, MI 48219

American Society for Testing and Materials
1916 Race Street
Philadelphia, PA 19103

Brick Institute of America
11490 Commerce Park Drive
Reston, VA 22091

National Concrete Masonry Association
2302 Horse Pen Road
P.O. Box 781
Herndon, VA 22070

Portland Cement Association
5420 Old Orchard Road
Skokie, IL 60077

9. Curtain Walls

James G. Howie

9.1 INTRODUCTION

Preventive maintenance (PM) with regard to curtain wall construction should be a part of the design drawings and specifications, should be reevaluated when shop drawings are reviewed, and should be discussed in depth during installation. The first principle is that an appropriate design with correct materials and knowledgeable installation is the best maintenance program. But "the ounce of prevention" also includes positive steps after the construction is in service. Preventive maintenance involves a series of direct actions to alleviate the shortcomings due to design and construction elements.

Curtain wall problems are often dramatic, and are often obvious to the world, not just to building owners and construction professionals. Curtain walls are used extensively on highrise buildings. Practical maintenance of an exterior wall of a highrise building presents a wholly different issue than maintenance of other building systems. First, the exterior wall of a highrise is usually a far greater proportion of the enclosure than a roof due to the building's height. Second, the exterior wall surfaces of highrises are difficult to inspect.

Periodic inspection of multistory vertical exterior surfaces is optimally done on a drop scaffold when a close visual survey is needed, but such a survey is a giant task, especially when the drop platforms have to be rigged. Most often surveys are accomplished with binoculars from below, or more optimally from an adjacent window when vision panels are operable. Obviously, the oblique view from a window limits the area of inspection to adjacent surfaces or conditions (Fig. 9-1). Maintenance for curtain walls is usually a reaction to various problems: air and/or water leakage, deformations of the "skin," spalled brick, cracked glass, bulging sealant, or simple aging/ deterioration of some of the pieces of the curtain wall.

Construction is putting a building together from a kit of parts. That the construction process is one of assembly should be obvious, but it is especially true for curtain wall. It is an assembly that must meet more and different

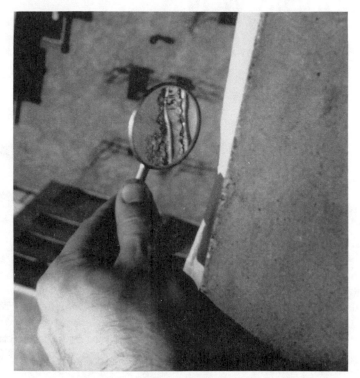

Figure 9-1. View from window with mirror—joint inspection.

criteria than other building systems, can be composed of different materials, and usually involves many tradespeople.

Basically, curtain wall is "in-fill"; it is not structural. Almost exclusively, the term refers to exterior wall construction, the "skin" enclosing openings between the structural parts of the building. Curtain wall is composed of glass, panels, frames, even masonry, and the necessary anchors of the assembly to the structure (Fig. 9-2).

The leaded glass windows (nonstructural exterior enclosures with aesthetic implications) of medieval cathedrals can be called curtain wall, as can cast iron framing in the nineteenth century (Paxton's Crystal Palace) and steel-framed industrial plants from the first half of the twentieth century. The emergence of framed curtain wall, as understood in contemporary terms, however, owes its existence to two post-World War II events: (1) the evolution of the multistory skeleton-framed building dictated by real estate demands and (2) the development of an aluminum industry with production facilities based on die-formed extrusions.

Although curtain wall includes various types of construction, this discus-

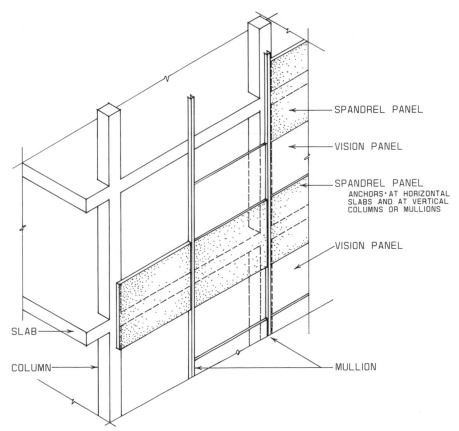

SPANDREL PANEL

VISION PANEL

SPANDREL PANEL
ANCHORS·AT HORIZONTAL
SLABS AND AT VERTICAL
COLUMNS OR MULLIONS

VISION PANEL

SLAB

COLUMN

MULLION

Figure 9-2. Curtain wall components.

sion will focus mainly on the aluminum and glass combinations that comprise the majority of installations. Masonry, wood, and concrete systems are addressed in other chapters. Although these systems were not dealt with as curtain walls, there are many similarities between a brick-bearing wall and a brick curtain wall.

In general, the materials of curtain wall construction are not problematic. The glass and other component materials are usually trouble-free and problems with curtain wall construction arise from a few root causes: design, product manufacture, and installation.

9.2 DESIGN

Good exterior wall design is the single most important factor in curtain wall construction and maintenance. In some instances, design may even overcome installation inadequacies. Design quality results from an understanding of how the system of a particular curtain wall will work, and from an acknowledgment of the forces that are going to act on the face of a building.

Exterior walls must resist an entire range of loads, and often have to resist them at the same time. Thermal stress, wind, rain, and ultraviolet effects are all deleterious and seldom act unilaterally. Other than structure, the most important loading, from a performance standpoint, is the pressure differential that all walls experience. The exterior wall must withstand a difference of pressure, between the exterior and the interior, across the thickness of the construction, whether it is a 12-in. solid masonry, a 6-in. stud-framed wall, or a 3–4-in. aluminum mullion glass-and-panel curtain wall (Fig. 9-3).

This pressure drop may be positive or negative; thus, windows can blow out as in Figure 9-4. The pressure will also vary on the surface of the exterior wall. A 10-story-high surface will experience different forces over the area, dependent on its siting, shape, and neighboring buildings. Indeed, the 2-story part of a 10-story facade will have different loads than a 2-story building alone, since the vortex of wind patterns will be different for these buildings.

A particularly important principle in exterior wall design from a technical viewpoint is redundancy or the "second line of defense." For example, in a masonry cavity wall (which is really a curtain wall), the system has an allowance for leakage: the flashing and weepholes. In a highrise metal and glass curtain wall construction, the thinking is similar when pressure equalization design (rain screen principle) is utilized, in that the line of defense is behind the exterior face of the wall (Fig. 9-5). The rain screen principle offers several advantages:

- The wind-driven water is kept primarily at the outside of the exterior surface of the curtain wall panel—that is, the outer surface of the construction.
- The watertight joint located on the interior surface of the panel experiences the pressure differential, with minimal exposure to moisture.
- The inside sealant joint is easier to inspect during installation and thus quality control can be enforced.

The goal of the rain screen principle is to isolate the joint and limit its function to withstanding pressure differential; the ultraviolet exposure and wind-driven rain are screened at the exterior face of the wall. Thus, the joint has to withstand only the pressure drop and physical movement.

Design procedure for customized curtain wall in a specific building involves model testing in a wind tunnel to establish the design load criteria. Then the curtain wall is engineered for the specific loading, and a mockup is produced for laboratory testing. The test procedures are well established by the American Society of Testing and Materials (ASTM), and the degree of loading indi-

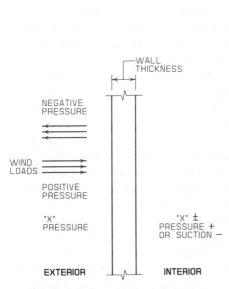

WALL THICKNESS

NEGATIVE PRESSURE

WIND LOADS

POSITIVE PRESSURE

"X" PRESSURE

"X" ± PRESSURE + OR SUCTION −

EXTERIOR   INTERIOR

Figure 9-3. Pressure drop across construction.

Figure 9-4. Window blow-out due to negative pressure.

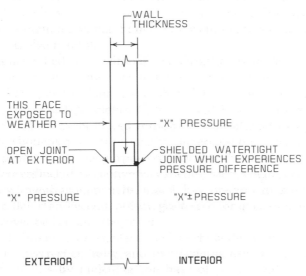

WALL THICKNESS

THIS FACE EXPOSED TO WEATHER

"X" PRESSURE

OPEN JOINT AT EXTERIOR

SHIELDED WATERTIGHT JOINT WHICH EXPERIENCES PRESSURE DIFFERENCE

"X" PRESSURE

"X"± PRESSURE

EXTERIOR   INTERIOR

Figure 9-5. Rain screen principle.

cated is for a specific design. However, the ASTM measurement criteria relate to individual pieces of the assembly—the window alone, or a specific manufactured element alone, say, the aluminum-framed glass and panel.

Prudent practice dictates that a full assembly mockup test progress from simple static testing, to cyclical testing, to full-scale dynamic testing. Static testing consists of applying a continuous pressure over a surface using a chamber or a box around the piece of construction that holds a difference in pressure. Cyclical testing consists of tests repeated in on/off cycles so that effects of multiple deformations can be assessed. Dynamic testing induces uneven pressure over a surface, produced usually by a propeller of an aircraft engine to the same pressure difference across the construction as static testing. Ideally, an on-site construction mock-up will then put all these pieces together for testing as a whole. In such a scenario, the mock-up will verify the design and engineering, the lab tests, and the installation, becoming the assembly quality control sample.

9.3   MANUFACTURE AND INSTALLATION

As the number and variety of applications has increased, the industry supplying aluminum curtain wall has expanded to meet the aesthetic needs of designers and the engineering needs of the product. A curtain wall product is usually composed of stock units of windows, glazed panels, vision panels, and so on from a manufacturer's inventory of pieces. The manufacturer/suppliers phase has several steps to it: (1) material verification, (2) shop drawings, and (3) sample approval.

1.   All products put into buildings have a paper history, which is documentation about the material quality (strength, wall thickness, etc.) and performance tests of the material and/or product. These descriptions are narrative information about the minimum specification particulars. For most installations, a supplier/manufacturer will try to work within a designer's profile of the exterior wall. But because there are as many different extrusions as manufacturers, a majority of curtain walls are adaptations of stock elements. Most manufacturers can accommodate particular design requirements with generic pieces (perhaps not *products* off the shelf, but often *extrusions* off the shelf). Extrusion dies are often modified to meet design intent and engineering requirements. These new formulations should be sample-tested (Fig. 9-6), since previous proven designs may not meet new criteria. The value of this testing is insurance that the part, be it mullion, panel, window, or major assembly, will perform adequately.

2.   Shop drawings are an important design and construction review. In forensic investigations of this nature, a basic understanding of the parts is necessary. With any kit, the instructions are critical. Instructions for curtain wall are the shop drawings primarily with the architectural or construction

Figure 9-6. Window testing.

documents as a requisite cross-reference. Performance data and catalog information that describe specific weatherstripping, sealants, cross-section of metal, and other materials are useful adjuncts.

Shop drawings give a view, usually full-size, of how the sections come together. These instructions are for use in the manufacturing process and are central to the installation coordination. Also, they have usually been reviewed by the professional for conformance to design intent. Shop drawings also show the relationship of other construction trades to a manufactured product that is to be installed. They describe critical aspects, such as tolerances, anchorage, and joint conditions. The interface of the shop drawings and the final product is a field-installed sample mockup, built for approval by the architect and owner.

3. Sample approval is the final review stage before construction of the walls begins. This phase consists of the full-scale assembly that contains the various pieces and provides an actual quality-control sample for product and installation. Field observation of the assembly is a principal quality-control measure. Assembly installation information offers specific procedures to be employed in the construction process. This information can also point to potential problems. For example, perhaps the top two floors were constructed during especially inclement weather (cold, wet, or hot); perhaps there was a change in personnel (different supervisor or inspector) or a different construction procedure was followed (windows at lower floor were lifted by outside crane, whereas those on upper floors were lifted in an inside elevator). Construction records are a resource about different construction practices that a contractor used. Although the exact nature of the problem is seldom evident, the clues from research about the construction can often be invaluable.

9.4   CURTAIN WALL FAILURE

Until recently, basic maintenance of exterior curtain wall has been principally concerned with window cleaning, but just as design and production have become more sophisticated, so has the scope of maintenance. Even so, curtain wall PM rarely exists as a program, perhaps because current systems on the market perform quite well; however, when problems appear, they can be significant. Unfortunately, there is no general solution, since there is no one common cause; fortunately, there are some applicable generalities.

Potential problem areas can encompass a few principal categories: (1) deformation of the "skin" (panel bulging, stress relief cracking, screw or rivet popping, and the like); (2) degradation of finishes (fading, peeling, delamination); and (3) leakage to interior (air and/or water).

1.   Evidence of deformation is the component failure and, as a result, the element has to be replaced. The cause of the problem also has to be resolved at the time of replacement; otherwise, without a problem cure, remedial action is of little value. There is no PM to avoid deformation. Rather, maintenance must focus on early identification of potential failure to avoid such problems. In Chicago and New York, surveys for early identification have been mandated by law; inspections by licensed professionals must be performed regularly, and evidence of problematic conditions requires notification of local authorities and a remedial program that includes repair or replacement, sidewalk protection, and a reinspection. Boston's Hancock Building is an example of a maintenance program that regularly monitors the walls (by field glasses) and replaces pieces of curtain wall, albeit as a result of significant failure and the attendant publicity.

Deformation problems manifest themselves in localized areas for the most part. The cause–effect usually has a clear relation since the brick spalls (Fig. 9-7), the panel buckles, the screws pop, and the glass cracks at the point of stress. Underlying causes can include structural or thermal movement, expansion due to moisture absorption, or simply misplaced shims, insufficient setting blocks, or screws that restrict expansion.

Also, when the glass panel cracks, the villain may be shims that have transferred some movement from the structure, or minor damage that occurred

Figure 9-7. Masonry spall.

in shipping or erection. The cause of these types of defects can generally be found in the vicinity of the defect itself. Since the problems are localized, they can be detected by

Expansion: Bulged sealant indicates joint movement. If joints are inadequate, stress occurs in the panel itself. Field cut pieces depend on the judgment of the worker and often have no provision for movement.
Moisture: Clay masonry materials expand readily from absorbed water. Mildew and/or stains indicate presence of water.
Different materials: The greater the mix of materials, the more allowance for movement is necessary.
Freeze-thaw syndrome: Water absorbed in a material will expand and cause the material to spall when moisture entry is uncontrolled.

Remedial action for deformation failures almost always involves extensive reconstruction, including removal and replacement of the damaged element. As mentioned, this replacement should obviously occur after the problem is assessed and a solution has been found.

2. Failure of the finishes, while easy to identify, is much more problematic. Finishes that delaminate are usually the result of inappropriate materials and complete replacement is the requisite repair. Colors that fade are generally products that are inadequate. Invariably, fading results from ultraviolet degradation, which has been understood for a long time. Finish performance testing is well established and a variety of products exist for different uses.

Finishes also degrade from atmospheric attack (air pollution, acid rain, high salinity, and the like). In this instance, the damage is more pervasive, since materials that are normally stable, such as aluminum and glass, are affected, and are often irreparably damaged. A specific maintenance effort in the face of external attack would be frequent wash-down of windows, panels, and framing. However, solutions to atmospheric problems are generally political not technical.

Finish failure due to aging alone is an obvious inevitability and maintenance will be of only limited value. Repairs to old finishes usually involve specialty coatings, which are exactly that — special materials and formulations, usually proprietary, and often installed by specialty contractors. In a vertically integrated example, KYNAR is a proprietary resin that is sold only to a limited number of companies, who then make coated products; then licensed contractors do the installation. The entire process is controlled and guaranteed by the original manufacturer. Such work has greater cost impact, but such cost will be more than offset by the value of a product guarantee. Remedial finish work is limited by product application requirements, and in many areas by environmental regulations. For example, Los Angeles and Orange County have strict limits on the allowable types of solvent coatings that can be used in an open environment.

3. The most problematic issue surrounding curtain wall failure is leakage — both air and water. Both affect the occupants of the building very directly; they are almost never caused at the point where the effect is seen, and both problems take a concerted effort to investigate. Air leakage to the interior is most often at the glazing and/or weatherstripping and therefore the perimeter should be checked first. Often weatherstripping in the reglet of an operating sash is stretched and then cut when installed; it then contracts, and gaps at the ends of weatherstripping are very common. In most cases, weatherstrip replacement, which is a simple maintenance task, easily cures this problem. Glazing is not as easily replaced, and the same type of gaps occur where rubber/vinyl pieces butt together; in addition, gaps appear at corners of the sash, where butted or wraparound glazing is used. Wraparound glazing (Fig. 9-8) is a U-shaped channel that slips over the edge of the panel and is partially cut, with a hinge so the 90-degree bend can be made. Wet sealant at the exterior is the only effective cure for air leakage of this sort. Assembly joints between panels, windows, mullions, and other elements can also be the culprits causing leakage and wet sealant application is an appropriate solution.

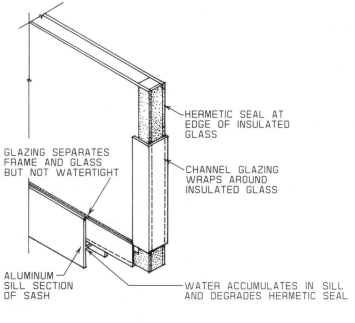

GLAZING SEPARATES
FRAME AND GLASS
BUT NOT WATERTIGHT

HERMETIC SEAL AT
EDGE OF INSULATED
GLASS

CHANNEL GLAZING
WRAPS AROUND
INSULATED GLASS

ALUMINUM
SILL SECTION
OF SASH

WATER ACCUMULATES IN SILL
AND DEGRADES HERMETIC SEAL

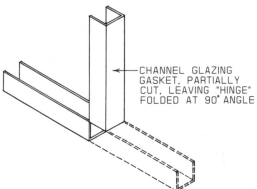

CHANNEL GLAZING
GASKET, PARTIALLY
CUT, LEAVING "HINGE"
FOLDED AT 90° ANGLE

Figure 9-8. Glazing, wraparound condition.

Water leakage is a much more bothersome failure both to the occupant of the building and to the owner. Seldom does evidence of leakage manifest itself where the problem occurs, so the area of investigation is greatly enlarged, and the possible causes are not readily apparent. Far too often, maintenance focuses on an application of sealant over every possible joint, when the cause might include just the opposite—blocked weephole drains or pressure equalization openings.

Preventive maintenance and remedial work are not generic. Each problem has a specific solution, and the repair efforts have to proceed from study to design to installation. Troubleshooting of a problem should cover the research (shop drawings, etc.) field inspection, and detailed specific investigation. At this point, the interior finishes will have to be removed and/or exterior inspection has to be accomplished by field glass survey, drop scaffold ride, or adjacent window inspection. The broad-scope categories of possible leakage problems include flashings, joints, sealants, and glazing.

9.4.1   Flashing

Flashing ranks first as a culprit relating to water leakage. Most curtain walls will have some sort of flashing, since the superstructure is essentially independent of the exterior skin and a flashing detail provides protection from any accumulated leakage from joints that have opened, from sealant that has failed, and from weepholes — since flashing works in concert with weepholes that will allow water in as well as out.

Where there is pressure equalization provision, there should always be a waterproof barrier (a flashing) toward the interior. It may be an integral part of curtain wall manufacture, rather than a field installation. If a water entry problem develops, flashing installed by different tradespersons, or in a different sequence than the curtain wall, should be a likely suspect. Flashing must be continuous, watertight to the inside surfaces, and proportionate to its expected performance; that is, the minimum vertical height is directly related to the wind load.

Flashing repair requires reworking the exterior wall, and almost always this work must proceed from the outside in. Design of such repair can be based on the construction document information, but will almost certainly need adjustment for existing conditions. As-built conditions can only be discovered during the progress of the work.

9.4.2   Joints

In construction, the "kit of parts" is assembled in less than ideal conditions; joint tolerances forgive these conditions and also allow for physical movement from the other forces exerted on a building, mainly thermal loads and physical movement from wind loads. Since joints experience significant movement, they should be another primary stop in a review of potential problem areas. Joints are potential trouble, since they are the location where the continuity of the exterior skin is interrupted. In some systems the joints will be closed with dry (usually shop-installed) weatherstripping, especially where the openings are operable.

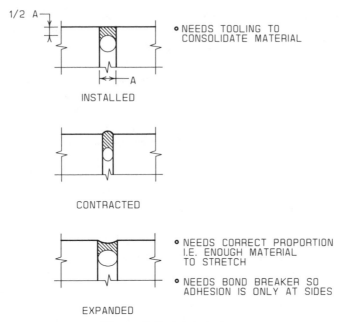

INSTALLED

o NEEDS TOOLING TO
CONSOLIDATE MATERIAL

CONTRACTED

o NEEDS CORRECT PROPORTION
I.E. ENOUGH MATERIAL
TO STRETCH

o NEEDS BOND BREAKER SO
ADHESION IS ONLY AT SIDES

EXPANDED

Figure 9-9. Joint design.

The major joint protection, however, in the current technology of construction is "wet" sealant, especially where the joint is between dissimilar materials. Assuming a good-quality product, a maintenance effort can address the sealant joints after an appropriate period, 10–15 years approximately. Failure before that point usually indicates some trouble with the design or installation.

Joints design does have basic criteria (Fig. 9-9) and for joints that have specific movement the installation can be evaluated. In other joints, especially shop-assembled joints, the evaluation is much more difficult. Such sealed joints rarely survive transport to site or erection and installation on the building. Interior treatment of these joints may be possible or perhaps some diversion of leakage to the flashing (in the case of water problems) can be provided.

9.4.3 Sealants

Remedial sealant work is a principal maintenance recourse, and material and product consideration is of prime importance. In general, the better the sealant (which is proportionately more expensive), the more successful the results. Replacement of a one-part acrylic sealant (which is good) with a two-part polyurethane (which is better) will increase the life of the joint assuming other installation aspects, such as compatible materials, clean and dry surfaces, appropriate primers, and of course, good workmanship, are equal. Sealant materials are complex chemical formulations and thus require some attention. Care should be taken with regard to using incompatible materials, multipart combinations needing quality control, or surfaces that may need preparation. Incompatible materials are frequent problems with sealant, since the chemistry base for each type of sealant is radically different. Polysufides will react poorly with acrylics, for example.

9.4.4   Glazing

The juncture of panels, either vision (glass) or solid, within a curtain wall, has a specific type of weatherstrip, or glazing. Because the glass panels have to be changed when broken, and because panel and frame move differently, glazing is used for separation of glass and frame. Wraparound vinyl is a common detail, due to ease of installation. Periodic inspection of glazing is a simple and effective way to monitor an important piece of wall to be followed by repair of obvious problems.

Glazing tapes are in common usage and are similar to wraparound vinyl glazing applications; likewise, wet sealant can also be used as a cure. Generally most manufacturers do not use both glazing tapes and wet sealant. Usually, only high-performance walls will have both systems. Glazing tapes are better protection against air and water leakage, but since they are not continuous, they must have a wet sealant at the butt joints in order to be fully watertight.

A common failure with insulated glass units is loss of the hermetic seal. This problem results from an accumulation of water in the sash if the glazing is not watertight, and the seal around the glass spacer deteriorates. A hermetic seal is the airtight protection for the insulation air between two panes of glass. The spacer for the layer of air usually has dessicant material to limit the humidity; when the seal breaks down from water exposure, the relative humidity cannot be controlled. The sash frames should have holes so that water can drain.

9.5   SUMMARY

Maintenance of curtain walls is a process of investigation, research, and corrective action. As with any other project, once the remedial or PM design decisions have been made, maintenance or repair construction must be done.

Maintenance and remedial procedures demand intensive and detailed inspection by professionals. Quality control during all phases of the construction and reconstruction is the principal means of ensuring effective PM, and although the problems may seem serious, the mysteries of curtain wall systems are solvable.

9.6   SUGGESTED READINGS

AAMA. 1988. *Window Selection Guide. Aluminum Curtain Wall Design Guide Manual.* Chicago, IL.

Allen, Edward. 1985. *Fundamentals of Building Construction.* New York: John Wiley & Sons.

Latta, J. K. 1973. *Wall Windows and Roof for the Canadian Climate.* Ottawa, Canada: Division of Building Research, National Research Council.

Maslow, Philip. 1974. *Chemical Materials for Construction.* Farmington, MI: Structures Publishing Company.

National Research Council of Canada. 1976. *Cracks, Movements and Joints in Buildings.* Ottawa, Canada.

Wilson, Forrest. 1984. *Building Materials Evaluation Handbook.* New York: Van Nostrand Reinhold Company, Inc.

9.7   ASSOCIATIONS WITH INFORMATION ON CURTAIN WALL CONSTRUCTION

American Architectural Manufacturers Association
2700 River Road
Des Plaines, IL 60018

Sealant Engineering & Associated Lines (SEAL) Association
7867 Convoy Court
San Diego, CA 92111

Sealed Insulating Glass Manufacturers Association
111 East Wacker Drive
Chicago, IL 60601

10. Grounds

Gerald R. Andrews

10.1 INTRODUCTION

A professionally designed, installed, and managed landscape grows in value each year. It is an investment that enhances image and creates positive first impressions of beauty, prestige, and practicality. An attractively designed and maintained property will:

- Contribute to a healthy and positive environment for employees
- Convey sound judgment and good taste to clients or customers
- Help to sell, lease, or rent the property
- Announce publicly that the property is well managed and the owner is a conscientious member of the community

Landscape maintenance is more than mowing grass. It starts with decisions made on the drawing board by the landscape architect and continues in the field by the landscape installer. Landscape maintenance is the art, skill, and knowledge of horticulture.

The basis of sound building maintenance management applies to landscape maintenance. But landscape maintenance differs from building maintenance since it deals with living things that grow and continuously respond to their environment. The landscape is always changing; it is never static.

Effective landscape maintenance represents a commitment by landscape professionals and owners. To ensure that grounds are attractively maintained, principles of cost-effective management must be understood and practiced. Individuals involved in landscape management should be dependable and cost-conscious practitioners as well as experts in the areas of trees, turf, and ornamentals. They must be skilled to perform their jobs efficiently and effectively by employing experienced, qualified, well-trained personnel and by using reliable, modern equipment. They should strive to keep the landscape looking its best within the confines of a predetermined budget.

10.2   GROUNDS DESIGN AND CONSTRUCTION: THE FIRST STEP IN GROUNDS MAINTENANCE

Landscape managers should always look for ways to provide quality within the established budget and time frame. This process begins at the design stage. Decisions made during design affect future maintenance. The choice and arrangement of plants and the way construction materials are specified dictate how much maintenance will be needed.

Avoid marginally hardy plants. Also avoid plants that litter. Specify plants that require little care and are tolerant of adverse conditions of the area, such as drought or salt air. Choose perennials over annuals because they require less care. Masses of plants and ground covers are easier to maintain than grass lawns. Beds should be mulched to discourage weed growth. Mulch also gives an attractive contrast to green plants and is less inviting to walk through than bare ground. Therefore, foot traffic damage can be minimized. Use sharp edges so lawn mowers can cut along a well-defined line. Some areas could be designed with a brick edging so that one wheel of the mower could run along it as the mower cuts. In general, minimize lawn areas, especially on steep slopes. Specify appropriate fence or barrier materials. For example, stone walls or stained wood require less maintenance than painted wood surfaces or hedges.

Raised beds are easier to maintain than plants at the ground level. Using a concrete pad for lawn elements, such as lamp poles, eliminates hand trimming. Arrange walks wisely according to how people move. Many walks have been designed that most people totally ignore in their desire to take the shortest route. Plan walks where traffic is expected and arrange for good drainage for the paths. If planting trees, carefully prepare the planting area, making sure to leave enough room for growth and development. Landscapers should remember to remove guide wires after the tree is established. Often these girdle the tree and can kill it. In a densely planted area, if shelter trees are removed from around a tree to be retained, the tree may fall due to the lack of support from other trees and its weak root system. Beware of trucks and other construction equipment driving over planted areas. Always space plants so that there is adequate room between plants for maintenance work.

10.2.1   Maintaining the Standards

The ultimate quality of a landscape is the result of the maintenance standards to which a company subscribes. To provide the framework for these standards, landscape maintenance philosophies must be examined prior to determining the levels of maintenance quality. Decisions are to be made on whether weeds are to be tolerated in beds, how crisp turf edges should be, what kind of pruning is acceptable, and height of the hedges. Standards should be established for site design, site construction, and horticulture practices.

Design considerations include choosing plants to depict a certain feeling or theme. Any deviations from the design concept should be reviewed in light of how changes impact the original design intent. Likewise, maintenance and pruning—or lack of it—can radically alter the ways plants function in relation to each other and to their surroundings. High standards for horticultural practices provide the basis for building a quality landscape. These standards should include the essential ingredients for healthy plant growth.

Once standards for design, construction, and horticultural practices are established, communicate them to those who will be required to follow them. The landscape should continuously be evaluated against the standards by regularly inspecting the site. A plan should be developed to ensure compliance with standards. Once the elements of the plan are implemented according to the established standards, continuous evaluation of the site assures that the standards are maintained.

Walk-throughs are a common method for evaluating landscape quality. A checklist outlines the standards, allows them to be clearly communicated, and serves as a valuable training tool (Figs. 10-1 and 10-2).

10.2.2   Estimating and Scheduling Landscape Maintenance

Webster defines estimating as "to form a rough judgment regarding the value, size, weight, degree, extent, quality, etc. . . . to calculate approximately." In practical business terms, estimating is used to embrace two areas: (1) estimating the amounts of labor (work-hours), and material and equipment necessary to complete a specific job, and (2) estimating the costs of labor and material. These two areas are then followed by scheduling (Fig. 10-3 on pages 276–277) the work so that the project can be completed within estimated costs and on time. Estimating addresses three questions: What is to be done? What quantities are required? How much time is needed to complete the project?

"What is to be done?" includes the project design requirements and the standards to be adhered to. These requirements should be clearly outlined so they can be easily followed by a landscape contractor or in-house staff responsible for landscape work.

"What quantities are required?" concerns how many acres of turf are to be maintained, how many trees, shrubs, and evergreens need care, and how many feet of curbs and walks need to be edged. The more detailed the knowledge of a site and its various elements, the more accurate the estimate will be.

(Text continues on page 278.)

LANDSCAPE STANDARDS

GENERAL

Standard	Exceeds Standard	Meets Standard	Needs Improvement
1. No dead or dying plant material visible in or from guest areas			
2. Landscape areas free of litter			
3. Visible irrigation components in good repair, valve box covers in place			
4. Paved areas free of weeds			
5. Lighting operational and not obstructed by foliage or mulch			
6. Hazards/work areas clearly marked for safety			
7. Work areas cleaned up prior to opening			
8. Tools and personal effects not in use out of guests' view			
9. Loud equipment not in operation after opening			
10. Labels and signs properly placed and maintained			
11. Posts, fences, and walls free of wheel marks, trimmer damage			
12. Costume/grooming standards evident and adhered to			
13. Unfinished work appears as complete as possible when unattended			
14. Design intent defined			

Comments:

ANNUALS

Standard	Exceeds Standard	Meets Standard	Needs Improvement
1. Weed free			
2. Good health and vigor			
3. Cultivated			
4. Appropriate for season and location			
5. Beds free of litter			
6. Damaged plants removed and replaced			
7. No flowers present on foliage crops			
8. Dead or yellow foliage removed where appropriate			
9. Pinched/dead-headed as appropriate for uniformity and to prolong bloom			

Comments:

Figure 10-1. Checklist for evaluating plants. *(Prepared for Disney World. Courtesy: Lied's Landscape Management.)*

LANDSCAPE STANDARDS

SHRUBS

Standard	Exceeds Standard	Meets Standard	Needs Improvement
1. Weed free			
2. Good health and vigor			
3. Sufficient mulch present, fresh in appearance			
4. Pruned to maintain appropriate size			
5. Pruned to avoid "coat hangers"			
6. Dead or dying branches removed			
7. Clippings removed and not visible			

Comments:

TREES AND PALMS

Standard	Exceeds Standard	Meets Standard	Needs Improvement
1. Pruned to appropriate size and effect			
2. Good health and vigor (no diseases or pests)			
3. No dead wood			
4. No cross branches			

	Exceeds Standard	Meets Standard	Needs Improvement
5. No weak crotches			
6. Limbed up to safe height			
7. Flower stalks and boots removed (palms)			
Comments:			

TURF

Standard	Exceeds Standard	Meets Standard	Needs Improvement
1. Weed free			
2. Good health and vigor			
3. Correct mowing height, frequency, and blade sharpness			
4. Crisp edging			
5. No scalping, uneven cuts, or ruts			
6. Appropriate use of string trimmer			
7. Clippings do not detract from appearance			
8. Paved areas free of clippings after mowing			
9. Newly sodded or repaired areas match existing turf			
Comments:			

Figure 10-1. (Continued)

273

PRODUCT	SIZE	QUANTITY
INSECTICIDES		
Chlordane		
Cygon		
Diazinon		
Dormant Spray		
Lime Sulphur		
Dormant Oil		
Home Orchard Spray		
Kelthane		
Lindane		
Malathion		
Metasystox		
Orthene		
Orthenex		
Pentac		
Sevin		
Slugit		
FUNGICIDE		
Benomyl/Benlate		
Bordeaux		
Captan		
Funginex		
Phaltan		
Sulphur		
Zineb		

PRODUCT	SIZE	QUANTITY
MISCELLANEOUS		
Compost		
Edging - Type		
Grass Seed - Type		
Growth Regulator Embark		
Mulch - Type		
Peat		
Plant Starter		
Potting Soil		
Pruning Paint		
Sod		
Straw		
Super Thrive		
Topsoil		
Typar		
LAWN SERVICE		
Fertilizer		

PRODUCT	SIZE	QUANTITY
TREE SERVICE		
Cable		
Cable Clamps		
Eyebolts		
Foam		
Lag Hooks		
Mauget Injections		
Medcaps		
Miscellaneous Hardware		
Pruning Paint		
Thimbels		
Tin		
Tree Rod		
Turnbuckle		
Washers		
FLOWER CARE		
Peter's Blossom Booster		
Three Way Rose Care		
GROUND COVER CARE		
PAVED AREAS		

FERTILIZER		Crab-X	Spike
Aluminum Sulphate		19-6-8	**WINTER PROTECTION**
Bonemeal		20-4-8	Boughs
Easy-Gro Packet		10-20-28	Burlap
Evergreen & Azalea Food (10-8-7 Ortho)		18-5-9	Hardware Cloth
Evergreen & Tree Food (12-4-8 Vertagreen)			Poison
Fertilizer Spikes		Herbicide	Repellants
Fish Emulsion		Mecomine D	Thiram
Gardeners' Special (11-15-11 Fertilome)		MSMA	Tree Wrap
Iron Sulphate		Tupersan	
Liquid Iron		Trimec	**PLANTING**
Milorganite			Hose
Miracid		Fungicide	Rope
Osmocote		Chipco 26019	Stakes - Type
Rapid-Gro			Strapping
Rose Food (Ortho)		Insecticide	
		Aspon	
HERBICIDES			
Fusalide			
Post		Other	
Roundup		Micronutrients	
Ronstar G		Soil Conditioner	
Surflan		Wetting Agent	
Treflund			
Tupersan			

Figure 10-2. Materials checklist (*Courtesy: Lied's Landscape Management.*)

275

Figure 10-3. Scheduling format. (Courtesy: Lied's Landscape Management.)

LIED'S

N63 W22039 Highway 74, Sussex, WI 53089 414/246-6901   ■   8616 Hwy. 45, Neenah, WI 54956 414/725-3877

DESCRIPTION OF WORK	DAILY	WEEKLY	MONTHLY	AS REQUIRED	COMMENTS
HORTICULTURIST SERVICE SITE INSPECTION					
DISEASE / INSECT CONTROL					
SEASONAL SPRAY PROGRAM					
TRIM & SHAPE					
PLANT BED CARE					
WEED CONTROL a) Hand Weeding					
b) Chemical					
EDGING BEDS					
CULTIVATE					
FLOWER CARE					
GROUNDCOVER CARE					
TREE WORK					
WATERING					
PAVED AREAS: Clean Off					
Weed Control					

TECHNICAL MAINTENANCE

LAWN CARE	
DEBRIS COLLECTION	
MOW & TRIM	a) Small b) Medium
	c) Large d) Rough Cut
EDGED PAVED AREAS	
FERTILIZE	
WEED CONTROL	
DISEASE / INSECT CONTROL	
AERATION	

SEASONAL	
THINNING & PRUNING	
SPRING CLEAN-UP	
FLOWER PLANTING	
FALL CLEAN-UP	
WINTER PROTECTION	

ADDITIONAL COMMENTS:

A number of methods can be employed for determining quantities. If a detailed as-built landscape plan is available, the use of a planimeter will allow the estimator to determine the exact acreage, areas of trees, shrubs, evergreens, and so forth. A planimeter is an instrument that will measure the square inches of an area.

If as-built plans cannot be obtained, aerial photographs available in scales as large as 1 in. to 50 ft. will give the necessary information. These photographs are generally available for purchase on file film for most urban area sites that are at least 2 years old. Check the yellow pages under Photographers–Aerial or contact the county assessor's office.

In addition to obtaining information by the use of planimeter or aerial photographs, it is still necessary to visit the site to assess the conditions of material and to verify the accuracy of the plan or aerial photographs. If as-built drawings or aerial photographs are not available, it will be necessary to physically measure the site and record the material found on the site.

The degree to which the components of a site are broken down will depend on several factors, including the complexity of the project and the approach the contractor chooses to follow.

As a minimum it will be essential to know:

1. Square feet of turf
2. Square feet of shrub beds
3. Square feet of flower beds
4. Number of trees and their average size

With this data, calculations can be made for the amount of fertilizer, herbicide, and fungicide required for the turf; the amount of fertilizer, herbicide, mulch, and insecticide required for the shrub beds; the number of flowers needed; and the amount of tree food and insecticides required. This makes it possible to get a rough comparison of maintenance costs of different projects on the basis of their common denominators.

Once what is to be accomplished and the quantities required are known, it is necessary to address the time allotment for maintenance tasks. Check manufacturers' information, but be aware that it may not be accurate. For example, an equipment manufacturer claims that a 21-in. hand mower will mow 6.79 acres in a day. Indeed it will, if a worker starts at point A and walks 32 miles to point B at the rate of 4 mph without stopping for 8 hours. This production standard is not reasonable.

In 1977, the American Landscape Contractors Association (ALCA) conducted a survey of maintenance contractors to develop production factors (Table 10-1). *Grounds Maintenance Magazine* published several similar surveys in the late 1960s and early 1970s. Equipment manufacturers frequently give cost comparisons using typical application methods and timetables. But it is necessary to conduct time studies and keep records to develop your own historical data.

Table 10-1.   Production Factors, ALCA Survey,
September 19, 1977

Operation	Quantity	High (hr)	Low (hr)	Mean (hr)
Mowing, hand	10,000 ft²	4.0	.5	2.0
Mowing, to 60 in.	10,000 ft²	1.5	.1	.4
Mowing, over 60 in.	10,000 ft²	1.5	.1	.25
Mowing, meadow	Acre	1.5	.5	.8
Spray weeds				
Small areas	10,000 ft²	2.5	.33	.5
Large areas	10,000 ft²	2.5	.18	.25
Edge curbs/walks	100 ft	1.0	.02	.10
Dethatch lawn	10,000 ft²	10.0	1.25	2.8
Weed shrub beds	1,000 ft²	2.5	.5	1.25
Prune shrub beds	1,000 ft²	8.0	.7	3.4
Shear hedges	100 ft	1.0	.08	.6
Prune groundcover	1,000 ft²	1.5	.25	1.1
Edge beds	100 ft	1.5	.25	.75
Leaf removal, beds	1,000 ft²	.5	.08	.28

In developing this data, define terms in writing so that all involved are using the same guidelines. Has trim time been included? Time for off-loading and reloading of equipment? Refueling? Removing excess clippings? Routine equipment service such as oil changing, checking lubrication, and tuning up engines should not be left out.

For greater efficiency in calculating equipment maintenance costs a number of factors can be combined and these combinations can be treated as a unit. For example, instead of calculating oil change, lubrication, and engine tune-up costs separately each time to provide an estimate, establish a unit cost for engine maintenance including all three engine maintenance requirements. Production factors are needed for all operations that are routinely performed, as well as for those that are only occasionally undertaken. A production factor list should include the following:

Turf
mow (for each size machine used)
fertilize (apply dry material) by hand and by large machine

spray (apply liquid chemical) by hand and by large machine
leaf removal, large and small areas
edge curbs and walks

Shrubs, Evergreens, and Groundcovers
cultivate, easy and difficult
prune, easy and difficult
shear hedges, large and small

fertilize (apply dry material)
spray insecticide and herbicide
edge

Trees and Specimen Evergreens
    fertilize, large and small   prune, large and small
    water, large and small   spray, large and small

Flowers
    plant   pinch
    fertilize   remove
    water

Miscellaneous
    travel   loading

With this information used as a basis, the list of operations can be expanded to meet specific needs. Once a list is completed for a project, the time needed to perform different maintenance tasks can be easily calculated.

> Example: Calculate time needed to mow 12,000 ft², 26 mowings per year, with 21-in. machine (includes trimming and cleanup)
>
> $$4000 \text{ ft}^2 = 12,000 \text{ ft}^2/3\text{hr}$$
> $$3 \text{ hr} \times 26 \text{ mowings/yr} = 78 \text{ hr/yr}$$

This calculation gives the cycle hours for mowing the lawn of this specific site per season.

Forms (Figs. 10-4, 10-5, and 10-6)* have been used by landscape management companies in their cost-estimating activities. This type of detailed estimating has three distinct advantages:

1. Production personnel know exactly what was estimated.
2. Crew supervisors have a planning tool because they know how long each function should take.
3. By tracking actual time versus estimated time, the manager can easily spot problem areas.

Both the cycle time and total time are given to the crew on the work order (Fig. 10-7). The crew's daily job card (Fig. 10-8) tracks the time and is compiled for comparison with the estimate. Of course, the work should always be periodically inspected to make sure that it complies with the established horticultural standards.

---

*Figures 10-4 through 10-9 appear at the end of the chapter.

10.2.3   Cultural Guidelines for Ornamental Plant Care

Plant culture deals with the practice of growing plants with an emphasis on improved plant development. Plant culture consists of a number of areas that are critically important for ensuring healthy and attractive plant growth. These areas include soil fertilization, watering, cultivation, pruning, insect and disease control, and plant protection during the winter months and against mechanical injury (Fig. 10-9).

10.2.3.1   Fertilization

Plants depend on various compounds and elements found in the soil and in gases carried in the air. To maintain an appropriate level of fertility in the soil a periodic addition of manure or commercial fertilizer often is necessary. After being dissolved in water, the nutrients contained in manure or commercial fertilizers are taken up by the roots and are distributed throughout the plant.

Manure, generally, is a much more complete source of plant food than any commercial fertilizer. It provides the plants the necessary trace elements in which commercial fertilizers are lacking, and unlike commercial fertilizers, it substantially improves the soil. Manure is an ideal source of needed elements. Because of its shortage, however, commercial fertilizers are used extensively with satisfactory results. These fertilizers contain three basic compounds: nitrogen, phosphoric acid, and potash.

Nitrogen is an effective foliage growth stimulant that can be manufactured as a chemical or found on such animal by-products as dried blood or bone meal. Phosphoric acid, obtained from various phosphates, is essential for vegetables and flowering plants, and potash is essential for roots.

The best time to root-fertilize trees is when the trees have leafed out. For established trees—sugar maple, beech, oak—use Milorganite, Fe chelate, or other similar fertilizers in dry or granular form, placed in holes around the base of the tree to prevent runoff. Other deciduous and evergreen trees should also receive dry fertilizer: Use 12–4–8 (indicating nitrogen–phosphoric acid–potash content) and Fe chelate placed in holes. Add aluminum sulfate to increase soil acidity when fertilizing evergreens. Use a complete fertilizer such as 15–4–5 and Fe chelate in solution if fertilizing with a soil needle or probe. For newly planted trees—sugar maple, oak, beech—use fish emulsion with a wetting agent as a soil drench over the top of the ball. For other deciduous and evergreen trees, use liquid 20–20–20 along with wetting agent such as a soil drench.

Apply 19–6–9 to evergreen shrubs in spring or fall at the rate of 2 lb/100 ft^2 and water the fertilizer in. Flowering deciduous shrubs should be fertilized using slow-release low-nitrogen fertilizer in spring or fall. The slow-release fertilizer provides long-lasting continuous release of elements. Low-nitrogen content of the fertilizer minimizes burning water in the fertilizer.

When fertilizing groundcovers such as pachysandra, euonymus, vinca, and lamium apply Milorganite or another 6–2–6 slow-release product at the rate of 4 lb/100 ft^2 over the bed area in spring or fall. In addition, these plants may require supplemental liquid fertilizer during the summer. Yellowing indicates the lack of nitrogen. Some areas may need compost top dress to make the soil colder and increase its shading capacity. Aegopodium groundcover is susceptible to spider mites. If an attack of spider mites occurs, aegopodium should be cut back in the summer. Use Milorganite or a liquid fertilizer after cutting back the plants.

Add peat moss and compost to perennial flower beds in spring or at planting time. It must be worked in to improve the soil. Manure can be used in place of compost. Milorganite or any other 6–2–6 fertilizer can be added also, if geraniums are not present, since they may be affected adversely by nitrogen. In prepared beds, which should include manure and compost, add a complete fertilizer (nitrogen, phosphorous, and potash) 10–20–26. Add peat moss if beds need friability. Osmocote can also be used. This fertilizer contains plastic-coated nitrogen granules that result in a highly slow-release fertilizer. In new beds, add peat compost and bone meal, which is high in phosphorus. Supplementary liquid 15–30–15 fertilizer should be applied to annuals, once a month through early fall to improve blossoming. Perennials need not be fertilized frequently if the initial bed was prepared properly with plenty of well-rotted manure compost. Dry fertilize these plants in early summer for obtaining attractive foliage with 10–10–10. Organic soil amendments should also be added in spring or fall.

Roses should be fertilized in spring with a granular rose food, such as Ortho rose food, and bone meal. Since roses prefer fairly dry soil, an alternating cycle of liquid and granular fertilizer should be applied to the roses two times per month until late summer. Then fertilizing should be stopped as any late new growth will not harden off (i.e., become cold tolerant) in time for winter. Liquid fertilizer provides quicker results and if used every other time will not cause soil to be too wet.

10.2.3.2   Cultivation

Cultivation of soil is an effective means of weed control, moisture conservation, and soil aeration. By suppressing or eliminating the growth of weeds, competition for nutrients between weeds and desirable plants is eliminated. Since cultivated soil is readily compactible, especially by rain, frequent cultivating is necessary.

Hand-held and power gardening implements can be used for soil cultivation. Use caution when working up soil around the base of the trees, especially those that exhibit a shallow root system, such as Norway maple and birch. With more and more airborne disease problems such as maple decline, cytospera, and collar rots, basal cultivation of trees should only be done with the utmost care and discretion.

Deciduous shrubs should have porous soil at their base throughout their lifetime. Cultivation by spading the base once in the fall and lightly through the growing season assures proper soil aeration. Since evergreen shrubs are extremely shallow-rooted, no cultivation is recommended at the base of evergreens.

Additives such as peat or organic compost should be used if the beds become hard despite cultivation practices. Beds should be cultivated a total of four times per season, in addition to the fall spading. Shrub beds cannot be cultivated if they are mulched. As the organic mulch is broken down, however, cultivating can be done prior to adding a new layer of mulch.

Groundcovers need light cultivation throughout their first two seasons of getting established. After that cultivating should not be done because new growth and rhizomes could be disrupted. Some groundcovers such as lamium need not be cultivated after their first season because the plants have grown together. Established groundcover beds may be cultivated to incorporate organic soil amendments if the plant population is sparse and the ground is hard or depleted of nutrients and organic matter. Avoid deep cultivation and spading beds because the plant crowns and underground establishment of new plants could be disrupted.

Flowers should be cultivated lightly to keep the ground loose and the weeds under control until the flowers have become large enough so there is no competition from weeds. Cultivate perennials throughout the season as needed, noting carefully where underground bulbs and nongrowing perennials are located. Cultivate roses regularly to keep the weeds down and aerate the soil. Again, deep cultivation is not needed for roses.

Cultivation of semihardy broad-leaf evergreens such as rhododendrons, azaleas, and beanberry should be shallow, carefully noting where the plants' roots are located. Cultivation of the bed and around the plants is used to keep weeds down, aerate the soil, and incorporate fertilizing and amendments (i.e., additives such as lime, manure, sulfur, compost, etc. used to improve soil).

10.2.3.3 Pruning, Disbudding

Pruning of plants is necessary for a number of reasons. A certain amount of growth of trees and shrubs needs to be cut back periodically to keep the plant from spreading beyond desired boundaries and for shaping it in order to fulfill esthetic requirements. Pruning of flowering trees and shrubs encourages production of attractive blossoms.

In general, pruning should be done when the plant is dormant. On flowering plants, prune after the blooming has ceased, early summer for spring-flowering plants, and late fall or winter for plants that bloom late in season.

Deciduous trees: Prune out as needed to correct aesthetic and potential structural problems. No more than one third of the plant should be removed at any one time. Prune in spring or when the trees are dormant, except in unusual situations such as storm damage, unusual growth spurts during the season, or if growth is interfering with traffic.

Fruit trees: Prune when dormant, taking into consideration the desired branching structure and the kind of fruit tree.

Pine trees: Pinch in their candle stage to keep the desired shape and form. Spruce and other evergreens can be sheared in spring before new growth hardens off or they can be pruned in late summer.

Deciduous shrubs: Pruning depends on flowering time, species of shrub, and age of shrub. Early-spring-flowering shrubs such as forsythia should not be pruned until after blooming. Lilacs should have flower seed heads removed immediately after blooming. Pruning too far below the spent seed heads will result in no flower bud formation or flowering for the following year. Other deciduous shrubs such as spirea should be cut back to within 6 to 8 in. of the ground in early spring. For such shrubs as dogwood, viburnum, honeysuckle, and ninebark, renewal pruning (one fifth to one third of wood at ground level) should be done in spring to ensure a full season for regeneration of new growth.

Evergreen shrubs: Shrubs such as junipers and Japanese yews need one or more prunings in a season to keep the desired shape. Mugo pines should have candles pinched before they develop into needles and harden off.

Euonymus: Prune with a hedge trimmer two or more times per season. Vinca can also be trimmed in the same manner.

Pachysandra: Trim along sidewalks if growth encroaches too much. They should not be trimmed other than to keep top growth uniform.

Aegopodium: Allow to grow back until midsummer or right after blooming, when new plants are forming at the base of the old ones. Prune the existing plants back to within 3 in. of the grade. New plants will generate from the established old root system, resulting in new healthy plants until winter or frost.

Lamium: It is an aggressively growing groundcover and should be trimmed in spring to keep plants within bounds. Prune as necessary during the season to achieve the desired effect.

Annuals: Pinch back growth on annuals such as mums until 7 or 8 weeks before frost date, in order to achieve a full plant. At planting time, some annuals, depending on growth rate, need to be pinched back to prevent legginess. Flowers such as petunias, marigolds, and dwarf dahlias should be dead-headed as needed, that is, spent blossoms should be removed.

Perennials: Perennial spent flower heads should be removed as soon as possible. Maturing foliage on stems and at the crowns of these also need to be removed periodically to minimize unsightly appearance.

Roses: Hybrid teas, floribundas, and grandifloras need to be pruned in spring after uncovering. Remove thin, weak canes leaving four to five main healthy stems. Since winter will kill the upper stems, pruning must be done just above a bud on live wood. The plant will probably be no higher than 12–18 in. tall after the first spring pruning. Subsequent seasonal pruning during the blooming season should be done as the flower buds begin to open if the blooms are for cutting, or prune as the bloom has reached its peak and begins to decline. Prune back to the first compound leaf, containing five leaflets, on strong stem structure. Spent flower heads should be left on roses after late summer so that the plant hardens off for winter.

Semihardy broadleaf evergreens: Plants such as Korean boxwood require little pruning except for normal trimming of hedges. Sunburn can occur on boxwood if too much foliage is removed, exposing the normally shaded plant parts. When pruning is needed, be sure one or more dormant buds are present below the pruning cut. Spring pruning is most desirable for broad-leaf evergreens.

10.2.3.4 Insect and Disease Control

In order to maintain an attractive landscape, at times some type of insect and pest control becomes a necessity. Insect control methods fall into three basic categories: mechanical, biological, and chemical.

Although mechanical and biological methods have proven to be successful to a degree, the most widely used insect control method involves chemicals. In recent years, concerns have arisen about environmental damage caused by insecticides and pesticides. When using chemicals, a high degree of caution must be exercised to prevent adverse effects on humans and the environment.

Overall, learn what species of plants have potentially recurring problems. Find out the types of insects that are causing the damage. With insect problems, learn the pest's life cycle so it can be treated quickly and correctly. It is critical that insects and disease problems are monitored and controlled on a continuous basis. If plants become diseased and their treatment is beyond PM objectives, contact experts for assistance.

Spraying when the weather is hot and dry can cause toxicity. Spraying should be done early in the morning and late evening when temperatures are cooler and winds are calmer.

For trees an ongoing preventive program, rather than a reaction to a problem, is the best approach when dealing with pests. Dormant spray—lime sulfur with oil—should be applied to all trunks and branches beginning in early spring when daytime temperatures remain at 40°F or above for 6–8 hr at a time. Use a fungicide, a quick-acting insecticide such as Malathion, and chemicals specifically suited for certain insects and diseases (i.e., use Pentac for mites, Dursban for borers, Diazinon for a very wide range of insect problems). Known outbreaks of specific problems such as leaf miner and birch borers should be treated routinely. Often it is advantageous to treat plants against a number of pests and diseases at the same time. Spruce gall aphid, bronze birch borer, zimmerman pine moth, apple scab, and birch leaf miner can be controlled in this manner. Healthy plant material has less chance of disease and insect infestation.

Shrubs can be treated in a very similar manner to trees. Specific outbreaks include honeysuckle aphid and aphids on viburnums. Groundcover plants should be sprayed whenever trees and shrubs are sprayed. Specific problems are pachysandra leaf blight, scale on pachysandra, and mites on aegopodium.

Flowers, perennials, and roses should be treated with combinations of insecticides, miticides, and fungicides every 2 weeks throughout the growing season. Changing chemicals, especially fungicides, is necessary to keep effective control of plant problems. Fungi develop resistance against certain fungicides if exposed to them over time.

Semihardy broad-leaf evergreens' main problems are usually collar rot, fungal leaf, and leaf miner. Drenching these evergreens with specific fungicides such as Subdue 2-E, Lesan, and Benomyl when problems start appearing is an effective control method. Careful selection of systemic insecticides is needed for leaf miner problems.

10.2.3.5   Watering

The plant material should be monitored for ground moisture during every visit to the job site. Most landscape contracts are written without including watering. However, if site inspection indicates that watering is needed, the client should be advised. Late summer irrigation especially has been reported to be an effective means of minimizing winter injury to ornamentals including evergreens. Providing adequate irrigation during this time is more effective than watering thoroughly just before the ground freezes.

10.2.3.6   Winter Protection

Plants possess different degrees of hardiness. Tropical plants, for example, are never planted in the northern zones unless their environment, temperature, and moisture are conducive to their growth, or unless the environment can be controlled. Greenhouses provide controlled environments.

Many plant types can withstand severe weather and may not require any protection. Some plants, however, need some protection from severe winters. When protecting plants against the winter, the objective should not be to keep the plants warm. The objective is to avoid excessive fluctuations in temperature and moisture depletion in the soil, to protect plants from severe drying from exposure to the winter sun, and to keep them cool so that new growth does not start too early in the spring. Plants growing too early in the season because of early warm spells are susceptible to damage by subsequent cold weather.

Lightly cover plants that are susceptible to winter damage to provide a stable environment during dormancy. For winter protection, wrap all young thin-barked trees, such as crabs, maples, lindens, ash, and honey locust, with wrapping tape until trees develop tough corky bark. Remove wrap in spring.

Guard against mouse and rabbit damage. Screen all mouse/rabbit-susceptible plants, especially shrubs, with hardware cloth. Plants susceptible to mouse/rabbit are crabapples, hawthorn, plums, cherry, burning bush, viburnums, serviceberry, willows, spirea, and dogwood. Severe winters mean rabbits and mice will chew most any plant. If it is impractical to use hardware cloth, use a thiram-based spray. Temperatures must be above freezing when spraying with thiram.

Apply antidesiccants such as Wilt-proof to evergreens that are just planted or planted in the previous season and to established evergreens having a history of winter desiccation problems. Late summer and fall evergreen plantings should be watered prior to winter. To prevent deer damage, hang a nylon hose containing human hair clippings in the tree or shrub. Spraying with thiram is also helpful.

Annuals are pulled in the fall so no winterization is needed. Perennials and perennial beds should be mulched in late fall, however, to minimize heaving. All dead aboveground plant tissue should be removed and destroyed to prevent disease/insect problems. Roses should be cut back to a height of about 30 in. Soil should be placed at the base of the plants to a depth of 12–18 in. Evergreen boughs or other protective material should then be placed over the plants primarily for the purpose of blocking the wind.

Groundcovers including euonymus varieties, pachysandra, and vinca should be covered with evergreen boughs, marsh hay, or straw to minimize heaving over winter. Bait stations with poison grain seed for mice should be placed in all groundcover beds under the cover.

Semihardy broad-leaf evergreens, such as rhododendron, azalea, mahonia, boxwood, wintercreeper, holly, and pyrancantha need to be protected over the winter months by mulching the base of plants with peat moss and wrapping individual plants to prevent moisture loss. Final appearance should resemble a small evergreen tree.

Water all evergreen trees to assure adequate soil moisture for the winter.

Note: Rodent bait should be used with caution on sites where children and pets are present. Always advise client of use.

10.3   LAWN CARE AND MOWING

Lawn mowing and trimming generally starts in spring and runs into late fall. A general rule of thumb when mowing is to never remove more than one third of the grass blade. Grass clippings can be left on the lawn after mowing if clumping does not occur.

The heights of mower cut will vary throughout the season with changes in temperature and moisture. In spring and fall when the grass is fast-growing, the height should range from 1½ in. to 2 in. In summer, when it is usually hot and dry, the grass should be left longer (2–3 in.) to provide some shading and to conserve moisture. The last mowing of the season should be short (1½ in. or shorter) to prevent diseases caused by grass blades laying over during the winter.

The frequency of mowing will also vary with the growth rate. Most lawns require weekly service. Rough or secondary areas are usually mowed less frequently. Mowing direction should be changed with each mowing when practical. When turf areas are continually mowed in one direction, the grass has a tendency to grow in that direction.

It is critical that mowing equipment be maintained in top operating condition. Check mower blades and blade housing decks daily. Decks should be free of debris and blades should be balanced and sharp. Change oil and grease according to schedule. Keep all mowing equipment clean to add to the life of the machine and to the professional image of the company performing the work. Trim edges with a line trimmer or shears to give a clean, sharp, final appearance.

Bluegrass lawns need 3–4 lb of nitrogen per 1000 ft^2/yr, depending on the release rate of the nitrogen used. Applications should be made approximately every 6 weeks.

Weed control is best achieved by spraying weeds at the end of summer. A follow-up application in spring will catch those weeds that have blown in. Amine forms of herbicide such as ester are recommended to prevent drift damage.

10.3.1   Lawn Watering

Proper watering is a vital part of a successful lawn-care program because 75–85% of the plant's weight is from water. Watering sustains growth during periods of inadequate rainfall and makes fertilizers and chemicals effective.

More watering is needed during hot, dry summer weather than during cool spring and fall weather. One good watering a week may be enough during cool weather, but several waterings a week may be needed during hot, dry weather.

It takes about 1 in. of water to soak the average soil to a sufficient depth. Place several straight-sided cans under your sprinkler and clock the amount of time it takes to fill them to a depth of 1 in.

Soak the soil to a depth of 6–8 in. when watering and then wait until the grass begins to turn a darker, blue-green color before watering again. "Footprinting" is another sign of moisture stress in cool-season grasses. Do not wait to water until the grass begins to wilt severely or turn brown.

If water runs off the soil surface before the intended amount has been applied, move the sprinkler to another location and finish that spot later. Avoid daily, light sprinkling because it encourages shallow roots, disease, weeds, and thatch. It also wastes a lot of water due to evaporation.

Clay soils and slopes make proper watering difficult. They get very dry and absorb water slowly. Water these areas with a sprinkler that applies water slowly. Be careful not to keep the soil saturated or the roots will die from lack of oxygen.

Morning is the best time to water, but if the grass needs water badly, water any time. If watering in the evening, finish watering 30 minutes before sunset so the grass can dry before night. Grass foliage that is wet during the night favors disease.

10.3.2 Dethatching

Thatch is a layer of dead organic material that gradually accumulates between soil and grass blades. It is composed of roots, stems, runners, and leaves. Excessive thatch (more than ½ in.) restricts movement of air, water, fertilizer, and pesticides into the soil. It favors insects, diseases, and shallow roots.

Infrequent mowing, mowing too tall, frequent watering, and growing grass on clay soils all contribute to thatch. When thatch accumulates to more than ½ in., it should be thinned with a machine designed for this purpose. Bluegrass and fescue should be dethatched in September, fertilized, and then overseeded. Zoysia and Bermuda grass should be dethatched in the spring before fertilizer and crabgrass preventors are applied.

10.3.3 Overseeding

Bluegrass, fescue, ryegrass, and mixtures of these grasses that have thinned out from summer heat stress can be overseeded in early fall to thicken the stand. If overseeding, weed killers used at that time should not be harmful to new grass seedlings. Spring seeding requires the use of a special crabgrass preventer such as Tupersan that is not harmful to new grass.

10.3.4   Aerating

Aerating helps especially high-trafficked thatched lawns and lawns growing on clay soil. The best aerating machines pull a round plug of thatch and soil from the lawn. The small holes in the lawn help the roots get oxygen, water, fertilizer, and chemicals into the soil. Aerate in spring or fall to relieve compaction. More severe problems may require twice-a-year aeration. The soil removed during aerations may be broken up and left on the surface to aid biological thatch decomposition.

10.4   OPERATING AND SERVICING LANDSCAPE MAINTENANCE EQUIPMENT

Grounds maintenance activities become increasingly sophisticated as more work is handled mechanically. Regardless of whether an outside contractor or inside staff performs grounds PM services, it is necessary to be well versed in the operation of maintenance equipment and have a good understanding of the latest landscape maintenance technology to run a cost-effective program.

Because special schools or union-sponsored training programs are almost nonexistent, in-house training is the main approach for upgrading the performance of equipment operators and training groundskeepers interested in becoming operators. A well-trained individual in equipment handling produces more efficiency, less downtime, less damage to property, safer operation, and prolonged equipment life. Establish a training program consisting of formal instruction, operation under the guidance of an experienced operator, a period of solo operation, an examination, and periodic refresher hints.

Operators should learn how to operate equipment as specified in the manufacturer's instructions, which is particularly important with new equipment and machines used seasonally. A good operator will know the limits of the machine, the danger involved with its use, and how to perform service functions such as greasing, changing oil, changing blades, and inflating tires.

Proper servicing keeps equipment at optimum working levels, reduces downtime, increases equipment life, and maintains trade-in value. The operators should be given a checklist of items for proper servicing of each type of equipment they use. Manufacturers' operating manuals usually include a list of items that require servicing and the frequency of servicing.

Checking for worn parts, frayed or broken belts, and other problems that could cause a breakdown or serious damage to the equipment should be included as part of the servicing operation. What to look for depends on the equipment. With experience, the trainee will become familiar with potential trouble areas and inspect them often. The trainees should use an operator's daily inspection guide and trouble report. A separate guide must be prepared for different types of equipment.

Operators also need to be acquainted with procedures involved in periodic inspection. The frequency is usually based on miles driven or hours of operation. An inspection form listing what and when to check reduces the chance of missing anything. The operators should know what needs to be done to winterize equipment for storage, and they should be able to place the equipment in service after storage.

10.5 LATEST COST-SAVING TECHNOLOGY IN LANDSCAPE MAINTENANCE

In recent years many technological developments have taken place related to landscape maintenance. New chemicals, more efficient machinery, as well as improved planting material itself, allow the grounds maintenance personnel to provide better services at lower costs and without any sacrifice to quality.

Since labor is the largest cost factor in grounds maintenance, reduction of work-hours on a project can result in substantial savings. However, as a labor-saving approach, some managers aggressively purchase labor-saving equipment and begin the widespread use of horticultural chemicals. It is common to overhear managers discussing success about a particular chemical or piece of equipment, only to find others who did the same with disappointing results.

There are several questions to consider when choosing and implementing labor-saving techniques:

1. Are potential cost savings accurately estimated?
2. Is the company able to deal with the latest technology?
3. Do the maintenance personnel accept and/or can they be trained to work using the latest techniques?
4. Is increased use of chemicals acceptable?
5. Is the company willing to take risks?

Quite often the evaluation of savings is much more difficult to measure than it would seem. For instance, mower X can mow 5 acres a day and mower Y can mow just 3 acres a day. They both cost $3,000; therefore, mower X should be more cost-efficient. First consider these factors:

1. Can the same operator run both mowers?
2. Perhaps mower X will not fit into the company's pickup and needs to be

transported by trailer. That entails purchasing a trailer, rigging the pickup to pull it, and teaching the operator how to back up a trailer.
3. Sometimes the time saved cannot be efficiently used, so instead of a realized savings, costs are increased.

The more thorough analysis any labor-saving idea is given, the more accurate its assessment.

10.5.1   Training

After evaluating and adopting a labor-saving idea, the work force must be trained to make it effective. Using preemergent herbicides to reduce hand weeding is the perfect example. Gardeners must be trained on proper site preparation (thorough weeding, proper moisture level), and on incorporating the application of herbicides in a timely fashion. When weeds emerge, gardeners must control the weeds through spot spraying, not cultivating and breaking the chemical barrier. Unless a technology is incorporated into an operation with accompanying training, the cost savings or labor savings will turn into an increased material cost or extra labor.

People, by nature, are uncomfortable with change. From time to time the work force may fight a change unless their prior support is solicited. Therefore, an educational process that benefits the workers is critical. Ask the operators to try the new equipment. Take note of their feelings about it. They will probably offer new and greater insight for consideration.

If time is not allocated to explain the benefits, the workers often view the change as making their work harder, reducing the hours they can work (a security threat), or as a way for the owner to gain higher profits. These negative perceptions must be addressed to ensure success.

Chemicals carry a bad connotation, and many people are anxious about their use. A crew that arrives wielding spraying rigs can cause panic and change the plan. Or simply, other people prefer cultivated beds to the hard, smooth surface that exists with preemergent herbicide programs.

Good communication and training are necessary to successfully implement new technology. There are two areas of particular concern with respect to risk. The first is safety. The more equipment used on a job, the greater the potential for equipment-related accidents. The second risk concerns the use of chemicals. There is a health risk to operators and the danger of mistakes that can cause damage to plants, which may ultimately result in the loss of a job and the diminishing of one's professional reputation.

The newest area of technology is growth-control chemicals. The labor saving here can be more significant than in any other area. Yet today, it is still an emerging science, and the results can be somewhat erratic.

It is recommended that the landscape manager proceed with caution into technologies until a product has been tested on local plant material in a given environment. Limit spray activities to areas of low visibility until the product and application confidence are reinforced.

10.5.2   Examining the Operation

When embarking upon cost-saving ventures, it is important to take a hard look at the company's operation and whether it can deal with the latest technological advances. Some difficult questions must be posed:

1. Do you have the supervision and training to make it work?
2. Do you have the skills and horticultural knowledge to make chemicals work?
3. Do you have a good PM program and a good shop to take advantage of equipment savings?
4. Will the owner accept the change in chemicals and/or equipment?
5. Will the staff accept the changes?

10.5.3   Can Technological Advances Reduce Costs?

While reduced costs of a service or its improved quality, or both, can generally be attributed to technological advances in a particular field, such is not always the case. The nylon cord trimmer, for example, introduced in the 1970s added a new dimension to maintaining landscapes but proved not to be a cost-saving device. When it first appeared on the market it provided instant labor savings, made work easier, and reduced back-related injuries. These circumstances encouraged people to demand more trimming. Property owners today prefer the appearance of trimmed areas to tall grass or open soil around sprinkler heads, flower beds, trees, and so on, and require that gardeners trim areas never trimmed before. The initial cost savings for trimming ushered in by the cord trimmer were great; however, in time this tool created a new esthetic standard that is costly to maintain. The innovation, therefore, did not turn out to be the cost- and labor-saver everyone thought it would be.

If not used discretely, devices that appear to be labor-saving innovations can actually increase costs. Of course, the final evaluation of labor-saving technologies involves results that need to be carefully monitored. If the job looks better or the costs are lower, or both, then the program has been a success. Performance improvement in almost any industry is contingent on knowing how to use technology and on sharing the knowledge gained with the individuals involved in that industry.

10.6   USING HORTICULTURAL CHEMICALS TO REDUCE LABOR

Horticultural chemicals offer the landscape manager a potential for increased labor savings if the intricacies of their use are mastered. There are three main areas of potential savings to consider: preemergence herbicides, postemergence herbicides to eliminate hand trimming, and growth-control chemicals.

10.6.1   Preemergence Herbicides

The effective use of preemergence herbicides such as Surflan or Devrinol in landscapes can save thousands of dollars per acre in maintenance costs. In order to affect the most savings, the manager must have broad knowledge in three general areas: herbicides, weed identification, and cultural practices associated with preemergence herbicides.

A common misunderstanding about preemergence herbicides is a belief that all herbicides control all weeds. Actually, there does not seem to be one herbicide that controls all, or even most, weeds. Each herbicide works best against specific families of weeds. Therefore, an applicator needs to anticipate which weeds are likely to be a problem and what time of year the weeds germinate. For instance, crabgrass germinates when soil temperatures reach the mid-60s. Therefore, unless the herbicide is applied slightly prior to this time, the seed will germinate and the herbicide will not do its job.

Often, to get good weed control, it is necessary to combine several herbicides to broaden the range of control. Even combining herbicides will not produce 100% control. Some weeds, such as spotted-surge, morning glory, and nutgrass are quite difficult to control regardless of the herbicide used. In addition, the type of plant material on which the preemergent herbicide is to be used is a limiting factor in herbicide selection.

Each herbicide has a varying degree of selectivity in regard to plant safety. Generally, the herbicides necessary to control existing weeds are not completely safe for the plant material present. Thus, the more knowledgeable operators must create a solution that will provide the best control. Also, chances might be taken in heavier soils that would not be considered in sandy soils that are low in organic matter.

Cultural practices also affect control. The best control can be achieved by thoroughly weeding the area prior to planting and not disturbing the soil (which would break the chemical barrier) after herbicide application. Then, weeds should be controlled with spot applications of a postemergence herbicide. Additionally, extra fertilizing can offset the stunting effects of preemergence herbicides. Proper watering after application is important.

10.6.2 Postemergence Herbicides to Eliminate Hand Trimming

Postemergence herbicides can be used to eliminate a substantial amount of hand labor. By spraying turf with a contact herbicide around sprinkler heads, trees, and other obstacles, most hand clipping or nylon cord trimming can be eliminated. One spray application can reduce costs by up to 75%.

However, timing is critical to the success of this approach in high-visibility landscapes. The herbicides need regular applications to prevent large brown areas in the turf. Monthly spraying is usually sufficient. If 6 weeks pass prior to respraying, too much growth may have to be burned back, resulting in unsightly dead grass that people will probably find objectionable. The use of turf dyes in the herbicide mix can offset this problem to a degree.

10.6.3 Growth-Control Chemicals

The third main area of savings is through growth-control chemicals. This field is relatively new and exciting. Not only do growth control chemicals reduce labor but they reduce debris and lower dumping costs that have become significant in many regions. In order to determine which growth-control chemicals to use, start with the manufacturer's representative in the area. The chemicals perform differently on plant material in different climates.

One application of growth-control chemical can prevent future growth for 2–4 months depending on the species. The challenge is to learn the proper time and rates of application. Some chemicals should be applied just as new buds break, whereas others work best if three or four leaves have emerged.

A critical consideration with these chemicals is the risk of plant damage. There have been instances where, after successfully using a material for a couple of years on a particular plant, damage occurs for no explainable reason. Also, occasionally a plant sprayed with a growth-control chemical is attacked by insects and takes months to recover.

Although there are some difficulties associated with these chemicals, the potential saving is significant. Proper training and constant efforts to stay current on research and the availability of new chemicals and new uses of old chemicals are essentials for modern maintenance operations. More important, successfully using horticultural chemicals to reduce weeding and for size-containment pruning can free up labor to do the little extras that truly produce a quality landscape.

10.7   IN-HOUSE VERSUS OUTSIDE MAINTENANCE CONTRACTS

In many cases, it may be beneficial to depend primarily on in-house staff for ground maintenance. At times, however, outside grounds maintenance contractors may have to be employed. Since the in-house staff is kept busy with routine maintenance, when heavy seasonal demands or unexpected work arises, the in-house maintenance staff can find the work overwhelming and may have to resort to outside contractors. In this case, it is generally found to be a good solution. When making a decision whether to perform grounds maintenance with an in-house staff or to contract this work to the outside agency, consider the following points.

1. Does the site require highly technically trained people?
2. Is an investment in expensive equipment necessary in order to perform proper maintenance?
3. Is additional labor frequently needed?

If the answer to these questions is yes, then look into the possibility of contracting out landscape maintenance work. When dealing with the outside contractor all costs are known prior to signing the landscape maintenance contract.

Capital expenditures for equipment and the need to occasionally hire additional labor force is eliminated if an outside contractor is hired. Also, a reputable contractor, through the application of sound business and management practices, can provide highly professional services. Constant competitive pressure will keep most professional landscape management contractors performing effectively. Contractors realize that they are not retained on the basis of tenure or seniority but on the basis of performance.

One disadvantage of working with outside contractors is the difficulty of getting the contractor to respond without much time delay to urgent landscape maintenance requirements. Some contractors may not be good landscape managers. It is important to select outside contractors carefully. Before selecting a landscape management contractor:

1. References from those who have used the contractor should be contacted.
2. The contractor's reputation should be reviewed.
3. The contractor's track record should be examined to determine level of experience in performing the work.

Is the contractor certified to apply insecticides? Does the contractor have insurance and is he or she bondable? Does the contractor employ people qualified to do the work? What is their level of expertise? Does their appearance reflect positively on the image to be portrayed with the landscape? Does the contractor have sufficient field and office staff?

The contractor should be required to submit a list of all available equipment designating backup support. Can the contractor deliver the required services on time? Before retaining an outside landscape contractor, be sure that the person is fair and equitable and has a real concern for clients and employees. In order to establish and maintain a mutually beneficial relationship, the contractor must have the ability to listen and understand the client's needs and priorities and to determine whether they can be met.

10.8   MAINTAINING PAVED AREAS

Most any property or building complex consists of pedestrian walkways, vehicular drives, and parking lots. The materials selected for construction of these areas generally depend on the required durability, the available budget, and the desired esthetics. Because of exposure to weather, these areas usually require frequent maintenance and repairs. Maintenance needed on paved areas varies with the severity of environmental influences and traffic loading. Maintaining these areas is more complex in cold climates than moderate climates.

As in the case of most any building system, the performance of walks, drives, and parking areas depends on the quality of original design, material selected, material installation, and maintenance. Drives and parking surfaces can be constructed of gravel, pavers, asphalt, or concrete. When installing walks, drives, or parking areas, certain critical requirements must be met. The first major concern is drainage. Poor drainage is the greatest cause of pavement failures. While surface water can cause substantial pavement damage, infiltration and subsurface water can cause serious cracking and settlement problems. Before starting construction of paved areas, an evaluation of the need for a subsurface drainage system should be made.

Surface water can be controlled by providing appropriate crowns and slopes. The water can be made to drain into open ditches that can be installed on one or both sides of roadways or around parking lots. Unless the parking lot is to be sloped to the sides, drains should be installed within the parking area and should be connected to storm sewers or low areas (where the accumulation of water would not affect the pavement). Minimum pavement slopes of 1½% to 2% are recommended.

Another critical factor in new construction as well as maintenance of parking areas is the provision of adequate base. The base is generally made up of gravel or crushed rock. For sidewalks, the base is generally not used where drainage is adequate.

Like the maintenance of any building system, effective maintenance of pavement starts with inspection and early detection of potential problems. If pavement areas are not inspected, and minor defects are not repaired, these defects could develop into major failures. A qualified inspector needs to check pavement surfaces regularly, especially in the spring, fall, and after substantial rainfall. When inspecting parking lots, drives, and walks, the following questions should be answered:

1. Does the pavement drain quickly and are there drains or ditches provided to carry away water?
2. Are the drains and culverts open?
3. Are there any signs of settlement?
4. Is the surface wear extensive with pronounced ruts and ripples?
5. Is there an indication of frost heave?
6. What type of cracking, if any, exists?
7. What is the cause of cracking?
8. Are there potholes and spalled areas?

For graveled roadways and parking areas, PM procedures are relatively simple. The roads can be regraded and additional gravel added as needed. Ensuring adequate crown is especially important for good drainage. For areas paved with masonry units there are two approaches. Where sand is used for base, the main concern is keeping areas even. If certain areas become uneven, they should be leveled by removing the pavers and leveling the sand base. The sand base must be confined under the pavers; otherwise, water will carry away the sand, causing settlement of the units. Where mortar is used for the base and joints, the maintenance problems are similar to those of asphalt and concrete surfaces discussed later.

By far the majority of paved areas are constructed of asphalt or concrete. If improperly constructed and/or maintained, these areas are susceptible to serious problems that can be very costly to repair. Surface ravelling and wear of asphalt is generally caused by its exposure to the environment. On asphalt surfaces, aging or poor compaction also causes surface wear. If the surface wear is not excessive, apply a seal coat to asphalt surfaces to prolong the life of the pavement. Where the wear is significant, an asphalt overlay may be required.

Rutting and potholes of drives is another defect that can be readily observed; it is generally caused by traffic and sometimes by the base settlement. A structural overlay or replacement of heavily damaged sections is the only solution. On asphalt surfaces, heavy traffic can also cause what is known as rippling. It can be rectified by grinding down the ripples and overlaying the ground area with asphalt.

A relatively prevalent problem that is expensive and difficult to correct is pavement settlement. Settlement is caused by poor drainage, inadequate base, inadequate soil-bearing capacity, or in some cases unanticipated heavy loading. Settlement and frost-heaving problems are serious because drainage and base construction must be altered before this problem is corrected. If the drainage problem or the base construction is not resolved, and the surface is repaired, the problem will generally recur in a relatively short period of time. If the problems due to settlement are not severe, patching may provide a temporary solution; however, in most cases, reconstruction of these areas will be necessary.

Another area of major failures, relating to both asphalt and concrete areas, is cracking. The types of cracks that are readily observable are across-the-road cracks often caused by temperature fluctuations and material shrinkage. Cracks parallel to the road are generally caused by temperature changes or frost heaving. Extensive random cracking or alligatoring of asphalt is usually caused by unstable base, poor drainage, or repeated heavy loading. Random cracking of concrete is generally caused by porous concrete. Although some cracking may not be detrimental to traffic using the paved areas, problems arise when water enters the cracks and gets to the substrate, weakening the pavement support and accelerating deterioration. Areas not draining properly are very susceptible to this problem. It is critical to perform regular inspection of paved area surfaces and seal any developing cracks without delay.

There are numerous compounds available for sealing cracks in asphalt or concrete. The materials used for repairs must be compatible with the existing material. The materials must adhere to the surfaces being repaired and remain plastic in all temperatures. For good adhesion, cracks that are being repaired should be dry and clean.

Other major pavement defects appear as scaling and spalling of concrete. Scaling is often caused by freeze–thaw action, especially if inadequately air-entrained concrete is used. It can also be caused by poor finishing or improper curing. Spalled areas are usually the results of porous concrete, especially on poorly drained surfaces. Minor scaling and spalling defects can be repaired by chipping out the failed sections and by patching them with new concrete. Asphalt overlays can also extend the life of poor concrete pavement. Areas of pavement containing extensive spalling generally need to be replaced.

In order to perform effective maintenance of paved areas or of any other systems related to buildings, it is most important that causes of deterioration, the behavior of materials used in original construction, and repairs are clearly understood by the maintenance personnel. The key cause of most failures of paved areas is improper drainage, deferred maintenance, or inadequate original construction. For the maintenance and repair of paved areas to be successful, the causes of failures must be eliminated. Regular inspection and early detection of potential problems is the basis for a good PM program.

10.9   SUMMARY

Although there are similarities between preventive maintenance of grounds and preventive maintenance of building systems, overall, PM of grounds calls for a different set of requirements than PM of any building system. Grounds maintenance staff has to deal with living and constantly changing material that is susceptible to a variety of degrading factors: Lack of appropriate nutrients in the soil, drought, insect attack, and diseases are just a few examples. Grounds maintenance is also complex because of the many different materials available: trees, shrubbery, groundcovers, flowers, and grasses. Each of these categories contains virtually hundreds of types of plants, each type requiring different conditions for its optimal development. Additionally, grounds maintenance includes the maintenance of drives and parking areas, snow removal, outdoor lighting, and a number of other tasks. Clearly, grounds maintenance personnel must have greatly diversified skills and knowledge in order to deal effectively with their varied but interconnected assignments.

Grounds personnel must know how to select appropriate planting materials for the climate and the soil. They must have knowledge of fertilization, cultivation of soil, planting and pruning, insect and disease control, weathering, winter protection, and maintenance of parking areas and lawns. Unless effectively managed, these PM efforts can be very costly. A variety of new chemicals and equipment for landscape maintenance are available for cost-reduction purposes, but individuals involved in the maintenance work must know how to use this material and equipment. Careless use of chemicals can be hazardous to the environment and human health. Purchasing lawn equipment without being able to use it to its full potential can also increase costs of maintenance.

Often the management of an organization has to decide whether to use the in-house staff for ground maintenance or to contract the work. There could be advantages both ways. A well-maintained landscape provides a setting for buildings, protects the environment, and presents a favorable image of the organization.

10.10   SUGGESTED READINGS

Conover, Herbert S. 1977. *Grounds Maintenance Handbook*, 3rd ed. New York, St. Louis, San Francisco: McGraw Hill, Inc.

Grounds Maintenance Forms and Job Descriptions. 1989. Cockeysville, MD: Professional Grounds Management Society.

Johnson, Warren T., and Howard H. Lyon. 1988. *Insects That Feed on Trees and Shrubs*, 2nd ed. Japan: Comstock Publishing Associates.

Pirone, P. P. 1978. *Diseases and Pests of Ornamental Plants*, 5th ed. New York: John Wiley and Sons, Inc.

Rathjens, Richard, and Roger Funk. 1985. *1985 Guide to Turf, Tree, and Ornamental Fertilization: Weeds, Trees & Turf*, pp. 40–50. Cleveland, OH: Harcourt, Brace, Jovanovich Publications.

Shigo, Alex. 1986. *A New Tree Biology.* Durham, NH: Shigo and Trees, Associates.

Sinclair, W., H. H. Lyons, and W. T. Johnson. 1987. *Diseases of Trees and Shrubs.* Japan: Comstock Publishing Associates.

Urban Phytonarian Handbook. 1983. Department of Agricultural Journalism. University of Wisconsin–Madison.

Whitcomb, Carl E. 1986. *Establishment and Maintenance of Landscape Plants.* Stillwater, OK: Lacebark Publications.

10.11 ASSOCIATIONS WITH INFORMATION ON GROUNDS MAINTENANCE

International Maintenance Institute
P.O. Box 1244
Appleton, WI 54912

National Institute on Park and Grounds Management
Box 1936
Appleton, WI 54913

Professional Grounds Maintenance Society
301 Galloway Avenue, Suite 1
Cockeysville, MD 21030

[*Note:* Figures 10-4 through 10-9 appear on the following pages.]

	JOB							

LIED'S Landscape Management

AREA and TASK	LABOR							
	measurement sq. ft.	÷	production standard sq. ft./hr.	=	time standard adjusted	×	# times performed	=	total time	×
Horticulturist inspection								
Disease/insect control								
Seasonal Spray Program								
Trim and shape								
Plant Bed Care								
Weed Control								
Hand weed								
Pre-emergent								
Post emergent								
Edge beds								
Cultivate								
Flower care								
Groundcover care								
Tree Work								
Water								
Paved Areas								
Clean off								
Weed control								
Debris hauling								
Travel								
Cycle hours								
Total hours								
Labor Cost								
Material Cost								
Grand Total								

Figure 10-4. Technical maintenance estimator's worksheet. *(Courtesy: Lied's Landscape Management.)*

DATE _____ PREPARED BY _____

rate	=	extended labor total	product	MATERIALS				TOTAL L and M
				X area/unit = cost/unit	X # apps. = cost/app.	÷ cost/bag = total cost	quantity	

LIED'S Landscape Management	JOB								
			LABOR						
AREA and TASK	measurement sq. ft.	÷	production standard sq. ft./hr.	=	time standard adjusted	×	# times performed	=	total time ×
Site inspection									
Debris Collection									
Mowing: Small									
Medium									
Large									
Rough									
Trimming									
Edge Paved Areas									
Clean up									
Travel									
Fertilize:									
Weed Control									
Disease/insect Control									
Aeration									
Travel									
Verti slice/slit seed									
Lawn renovation									
Thatch									
Clean up									
Debris hauling									
Travel									
Cycle hours									
Total hours									
Labor Cost									
Material Cost									
Grand Total									

Figure 10-5. Lawn-care estimator's worksheet. *(Courtesy: Lied's Landscape Management.)*

rate	=	extended labor total	product	MATERIALS				TOTAL L and M
				X area/unit =	X # apps. =	÷ cost/bag =		
				cost/unit	cost/app.	total cost	quantity	

AREA and TASK	measurement sq. ft.	÷	production standard sq. ft./hr.	=	time standard adjusted	X	# times performed	=	total time	X
LABOR										
Cleanup										
Prune										
Vacuum Debris - Beds										
Turf										
Edge - Paved										
Beds										
Cultivate										
Fertilize - Trees										
Shrubs										
Gr. Cover										
Fl. Beds										
Weeding										
Pre-emergent										
Paved Areas										
Debris Hauling										
Travel										
Sub Total										
Pruning										
Travel										
Seasonal Color										
Annuals										
Roses										
Bulbs										
Travel										
Winter Protection										
Travel										
Cycle hours										
Total hours										
Labor Cost										
Material Cost										
Grand Total										

LIED'S Landscape Management

JOB _____

Figure 10-6. Seasonal/special task estimator's worksheet. *(Courtesy: Lied's Landscape Management.)*

DATE _____ PREPARED BY _____

rate	=	extended labor total	product	MATERIALS				TOTAL L and M
				X area/unit =	X # apps. =	÷ cost/bag =		
				cost/unit	cost/app.	total cost	quantity	
			Anti desiccant					
			Boughs					
			Hardware Cloth					
			Repellents					
			Poison					
			Burlap					
			Tree Wrap					

Figure 10-7. Work order. (Courtesy: Lied's Landscape Management.)

LIED'S ■

N63 W22039 Highway 74, Sussex, WI 53089   414/246-6901   8616 Hwy. 45, Neenah, WI 54956   414/725-3877

NAME _____

ADDRESS _____   DATE _____

CITY _____   HOME PHONE _____

   WORK PHONE _____

Production Report #		Work Order #		Complete? Yes No	Contract or T&M	Time Arrived:	Left:

FOREMAN: _____   CREW: _____   Manager Review _____

TECHNICAL MAINTENANCE	Cycle Time	# Times Performed	Total Hours	LAWN CARE	Cycle Time	# Times Performed	Total Hours
Site Inspection				Debris Collection			
Disease & Insect Spray				Mowing: Small			
Seasonal Spray Program				Medium			
Trim & Shape				Large			
Bed Care				Rough			
Weed Control: Hand				Trimming			
Chemical				Edge Paved Areas			
Edge Beds				Fertilize: Manual			
Cultivate				Tractor			
Vacuum Beds				Weed Control: Manual			
Flower Care				Tractor			
				Disease/Insect Control			
				Aeration			

Groundcover Care

Tree Work

Watering
Paved Areas: Clean Off
Weed Control
Thinning & Pruning
Planting

Sub Total

Verti Slice/Slit Seed
Lawn Renovation
Vacuum Lawn
OTHER
Winter Protection
Cleanup:
Debris Hauling
Extras:

Travel Allocation *
*Correct amount added
at end of day

Sub Total

MATERIALS	QUANTITY	@	=	AMOUNT

COMMENTS:

Figure 10-8. Format for tracking time spent on various landscape maintenance tasks. (Courtesy: Lied's Landscape Management.)

LIED'S

N63 W22039 Highway 74, Sussex, WI 53089   414/246-6901   ■   8616 Hwy. 45, Neenah, WI 54956   414/725-3877

NAME _____   DATE _____

ADDRESS _____   HOME PHONE _____

CITY _____   WORK PHONE _____

Production Report #   № 2001   Work Order # _____   Complete? Yes ___ No ___   Contract ___ or T&M ___   Time Arrived: _____   Left: _____

FOREMAN: _____   CREW: _____   Manager Review _____

TECHNICAL MAINTENANCE	MAN HOURS	@ RATE =	AMOUNT	LAWN CARE	MAN HOURS	@ RATE =	AMOUNT	TIME & MATERIAL
Site Inspection				Debris Collection				Technical Maintenance _____
Disease & Insect Spray				Mowing: Small				
Seasonal Spray Program				Medium				Lawn Care & Other _____
Trim & Shape				Large				
Bed Care				Rough				Total Labor _____
Weed Control: Hand				Trimming				
Chemical				Edge Paved Areas				Total Material _____
Edge Beds				Fertilize: Manual				
Cultivate				Tractor				Subtotal _____
Vacuum Beds				Weed Control: Manual				Tax _____
Flower Care				Tractor				
				Disease/Insect Control				TOTAL INVOICE _____

	Aeration				**CONTRACT**
Groundcover Care	Verti Slice/Slit Seed				Amount
	Lawn Renovation				Extras:
Tree Work	Vacuum Lawn				
	OTHER				Labor
	Winter Protection				Material
	Cleanup:				Subtotal
Watering	Debris Hauling				
Paved Areas: Clean Off	Extras:				Tax
Weed Control					
Thinning & Pruning					TOTAL INVOICE
Planting	Travel Allocation*				
	*Correct amount added at end of day				
Sub Total	Sub Total				

MATERIALS	QUANTITY	@	=	AMOUNT
COMMENTS:				
	Foreman's Signature			
TOTAL MATERIAL	Client's Signature			

311

TURF AREAS —		Jan	Feb	Mar	Apr	May	Jun	Jul	Aug	Sep	Oct	Nov	Dec
Soil aeration	Power raking				●								
	Core aerification				●	●	○	○	○	●	●		
Planting	Fall seeding preferred				●	●	○	○	○	●	●		
Watering	1" of water per week			○	○	●	●	●	●	●	○		
Mowing	Fine turf: mow at 3" down to 2" weekly			○	●	●	●	●	●	●	○		
Fertilization	4 applications per year				●	●		●	●		●		
	Dormant feed											●	
Weed control	Broadleaf comprehensive treatment					●	●		●	●			
	Spot treatment as needed						○	○	○	○	○	○	
	Annual weeds—pre-emergent			●	●	●							
Lawn renovation	Combined effects of aerate, overseed, water, fertilize			○	●	○				○	●	●	
Pest control	Majority of problems will occur within this time						○	○	●	●	○		
Disease control	Both preventive and corrective					●	●	●	●	●			
Gypsum turf	Salt damage control			●								●	
Edging/trimming	With each mowing					●	●	●	●	●	●		
Debris	Spring & Fall clean-up and every mowing					●	○	○	○	○	●		
Soil penetrant	With micro nutrients					●	●		●	●			
WOODY PLANTS: TREES, SHRUBS, GROUND COVERS													
Planting	Spring planting season — bare root				●	●							
	Summer/Fall — B&B, container							●	●	●	●	●	
	Winter — trees	●	●										
Weeding	Pre-emergents				●	●							
	Pull weeds						●	●					
Fertilizing	No applications during August & September				●	●	●	●			●		
Water	Check soil moisture. Deep watering 18"-24" if needed					○	○	○	○	○	○		
Disease control	Majority of problems occur within this period					○	●	●	●	○			
Pest control	Majority of problems occur within this period					○	○	●	●	●	○		
Plant repair	Immediately after damage by wind, ice, equipment, etc.	○	○	○	○	○	○	○	○	○	○	○	○
Pruning	Evergreens						●	●			●	●	
	Deciduous trees and shrubs	●	●	●	○	○	○	○	○	○	○	●	●
	Remove spent flowers						●						
	Ground covers						●		●	●			
	Renewal pruning	●	●	●							●	●	●
FLOWERS AND BULBS —													
Cultivation & soil amendments	Soil preparation for annual flowers and perennials					●				●			
	Soil preparation for bulbs								●	●			
Plantings	Annuals					●	●						
	Divide perennials					●				●	●		
	Flowering bulbs									●	●		
Insect & disease control	Majority of problems during this time period						○	●	●	●	○	○	
Fertilize	Every two weeks						○	○	○	○	○		
Remove spent flowers	Weekly						○	○	○	○	○		
SEASONAL —													
Remulch	Shredded bark			●	●								
Seasonal clean-up	Turf and planting beds			●	●	●				●	●	●	
Winter protection	Install											●	
	Remove			●	●								
	Wilt-Pruf yews										●	●	
	Thiram spraying	●		●								●	
Snow plowing	Plowing and ice control	●	●	○							○	●	●

Figure 10-9. Landscape management calendar. Open circles designate weather-dependent activities. Closed circles designate activities common for all areas of the country, regardless of weather. *(Courtesy: Lied's Landscape Management.)*

Contributors

Gerald R. Andrews
Owner/President
Landscape Associates
Neenah, Wisconsin

Mr. Andrews holds B.S. degrees in resources management and horticulture from the University of Wisconsin–Madison. He became involved in property maintenance management in 1974, and later he developed a maintenance management system that was adopted nationally by various organizations.

Before starting his consulting firm, he held a position as branch manager with Lied's—Landscape Management, Landscape Design and Development in Neenah, Wisconsin. Earlier he served as manager of Lied's Fox Valley Branch where he was in charge of the landscape management division.

He is a member of Professional Grounds Maintenance, the American Landscape Contractors Association Management Division, the International Society of Arboriculture, and the Wisconsin Turfgrass Association.

Rene M. Dupuis
President
Structural Research, Inc.
Middleton, Wisconsin

A leading international authority on single-ply roofing systems, Rene Dupuis obtained the B.S., M.S., and Ph.D. degrees in civil engineering from the University of Wisconsin–Madison. After graduation he worked for the National Science Foundation and served as an assistant professor at the State University of New York–Buffalo.

Since 1974, Dr. Dupuis has been involved in materials research, with emphasis on built-up roofing and single-ply systems. He has written and presented many articles on roofing technology and has conducted numerous investigations for building owners and architects.

Dr. Dupuis is a member of the American Society for Testing and Materials, the Construction Specifications Institute, the Single Ply Roofing Institute, and the American Society of Civil Engineers. He has participated in numerous roofing conferences throughout the United States, and in the International Symposium on Roofing Technology held in Brighton, England, in 1981.

William C. Feist
Project Leader, Wood Surface Chemistry
  and Protection
Forest Products Laboratory
United States Department of Agriculture
Forest Service
Madison, Wisconsin

William Feist holds the B.S. degree in chemistry from Hamline University, St. Paul, Minnesota and the Ph.D. degree in organic chemistry from the University of Colorado, Boulder.

He served as a research chemist for Esso Research and Engineering Company (now Exxon) in Linden, New Jersey, from 1960 to 1964. In 1964, he joined the research staff of the Forest Products Laboratory as a research chemist and became a project leader in 1983. His areas of interest include wood weathering and finishing.

William Feist is a member of the American Chemical Society, the Society of Wood Science and Technology, the Forest Products Research Society, and a number of other national and international societies. He is an author or coauthor of 140 publications.

Joan C. Freitag
Technical Writer/Editor
Springfield, Illinois

Joan Freitag is an experienced marketing consultant, public relations counselor, journalist, technical writer, newsletter editor, and publicist. For three years she has written and edited *Building*, the University of Wisconsin's Department of Engineering Professional Development newsletter for architects, contractors, and engineers.

Since June 1984, she has owned and operated her own public relations writing firm, Communication Specialists. In addition to editing the university newsletter, she has done work for Professional Communications, Inc., the nation's largest publisher of marketing information for health-care professionals; several hospitals; Employers Association of Greater Wisconsin; and Giddings & Lewis, a machine tool company. Prior to starting her own firm, she worked as a newspaper reporter and host for a radio program aired on WHA, a public radio station in Madison, Wisconsin.

Joan Freitag received the B.S. degree in journalism from the University of Wisconsin–Madison.

James G. Howie
Principal
Howie, Freireich and Gardner
New York, New York

James Howie holds the B.Arch. degree from the University of Detroit and has been involved in architectural practice with various firms since 1965.

Prior to establishing his firm, he served as Chief of Quality Control at the New York State Urban Development Corporation where he was responsible for the technical review of building systems and components and for the evaluation of building materials. In 1977, he formed a partnership with Andrew Freireich; in 1987, Susan Gardner joined the firm.

With his firm Mr. Howie is responsible for the management of all phases of project development including design, contract documents, and construction administration. The firm's project types include housing, commercial office space and interior design, institutional facilities, landmark buildings, and retail establishments.

Mr. Howie has served on the board of directors of the New York Society of Architects, and is a former president of the Architects Council of New York City. Currently he serves on the Industry Advisory Committee at the New York City Department of Buildings and teaches in the School of Architecture at Pratt Institute.

Albert W. Isberner
Consulting Materials Engineer
Portage, Wisconsin

Mr. Isberner received the B.S. degree in civil engineering from the University of Wisconsin in 1953. Since 1981, he has been self-employed as a consulting engineer to the concrete, masonry, and portland cement plastering industries. From 1979 to 1981, he worked as quality assurance manager for Construction Technology Laboratories of the Portland Cement Association.

His earlier-held positions include those of senior and principal research engineer with the Portland Cement Association in the Concrete Materials Research Department; senior applications research manager for the American Cement Corporation; and associate and research engineer for the Portland Cement Association in the Applied Research Section. He also worked for the U.S. Army Corps of Engineers and for the Wisconsin State Highway Commission.

Mr. Isberner is a registered professional engineer in the state of Wisconsin and is a qualified and certified lead auditor, ASME/ACI Nuclear Code. He is a fellow of the American Society for Testing and Materials and the American Concrete Institute. He was the recipient of the Walter Voss Award, ASTM, in 1985.

Lynn R. Lauersdorf
Project Manager
Bureau of Architecture
Division of State Facilities Management
Madison, Wisconsin

Mr. Lauersdorf, a registered professional engineer, holds the B.S. degree in civil engineering and the M.S. degree in structural engineering from the University of Wisconsin–Madison.

Since 1962 he has been employed by the state of Wisconsin where he serves as a specialist in masonry construction and in related areas. In this capacity, he has developed the minimum requirements for design, materials, and construction for new State of Wisconsin masonry buildings. Mr. Lauersdorf is responsible for a preliminary draft of the masonry section of the Wisconsin Building Code and serves as a consultant for its updating.

He is a frequent lecturer in construction seminars and an author of numerous publications on masonry construction. His latest contribution in this area is the Appendix to ASTM C-270, "Standard Specification for Mortar for Unit Masonry," published in the *Annual Book of ASTM Standards*, 1989. He is a member of the American Society for Testing and Materials, the National Society of Professional Engineers, the Wisconsin Society of Professional Engineers, the American Society of Civil Engineers, and the Masonry Society.

Len E. Lewandowski
Senior Project Manager
Computerized Structural Design, Inc.
Milwaukee, Wisconsin

Mr. Lewandowski holds the B.S. degree in engineering and applied science and the M.S. degree in structural design from the University of Wisconsin–Milwaukee.

Since joining Computerized Structural Design in 1986, he has been involved primarily in the investigative work of existing buildings and in the analysis and design of new commercial and industrial facilities.

Formerly, Mr. Lewandowski worked for Inland Steel Company for 16 years. There he held a variety of technical positions including that of manager of engineering of the building system's division. The division generated $50 million in gross sales for 45,000 tons of manufactured construction products annually. While at Inland Steel, he served as chairman for the Metal Building Manufacturers' Association–Construction Committee.

Mr. Lewandowski is a registered professional engineer in eight states and a former member of the National Society of Professional Engineers and the Wisconsin Society of Professional Engineers.

Raymond C. Matulionis
Program Director
Department of Engineering Professional Development
College of Engineering
University of Wisconsin–Madison

Raymond Matulionis is a registered architect and holds B.Arch., M.Arch., and Ph.D. degrees. At the University of Wisconsin he is responsible for developing and coordinating seminars in building design, urban planning, and building construction, including programs on preventive maintenance of buildings.

He is a professor in the College of Engineering and teaches building design and construction in the Curriculum for Construction Administration program. He is also director of the *Building* newsletter. For more than 15 years, Professor Matulionis has been providing numerous Midwest architectural firms with consultation and support and is responsible for the design of a number of award-winning projects including Madison General Hospital Laboratory Building; Industrial Education Agriculture Technology Building, University of Wisconsin–Platteville; and Reedsburg Technical College.

He is a member of the American Institute of Architects and former president of the southwest chapter of the Wisconsin Society of Architects.

Stephen R. Mulvihill
Managing Partner
ARCON Associates, Inc.
Lombard, Illinois

Mr. Mulvihill holds the M.B.A. degree from the Graduate School of Business, University of Chicago, and the B.Arch. degree from the University of Notre Dame. The founder of ARCON Associates, Inc., he presently serves with his firm as principal and consultant on energy and maintenance issues concerning institutional and commercial buildings. He has been responsible for managing and preparing energy audits and studies and preventive maintenance programs for more than 250 buildings, including offices, hospitals, airports, schools, universities, research laboratories, country clubs, warehouses, condominiums, highrise buildings, and exhibit galleries.

Prior to forming ARCON Associates, he managed the largest single energy conservation contract undertaken by the U.S. Navy covering 102 buildings at the Great Lakes Naval Base. Mr. Mulvihill's studies for the U.S. Navy's Energy Conservation Investment Program established the procedure standards for consulting architects and engineers employed by the Northern Division of the Naval Facilities Engineering Command in Philadelphia.

A registered architect in Illinois and Michigan, he is a member of the American Institute of Architects and the Association of Energy Engineers.

Arthur L. "Pete" Simmons
Consultant and President
Roofing Consultants, Inc.
Los Angeles, California

Mr. Simmons holds the B.C.S. degree and is a certified roofing consultant. He founded his roofing and waterproofing firm in 1969. His experience includes positions of executive director, the Roofing Contractors Association; executive manager, the Roofing Inspection and Consulting Service; administrator, the Roofing Industry Improvement and Promotional Trust Fund; and owner/manager of his roofing/sheet metal contracting business. He is a member of the American Society for Testing and Materials, the Construction Specifications Institute, and the Roof Consultants Institute. He is also an associate member of the National Roofing Contractors Association and faculty member of the Roofing Industry Education Institute.

Mr. Simmons is a frequent speaker and lecturer at seminars and training sessions for organizations including the American Institute of Architects, the Construction Specification Institute, the National Roofing Contractors Association, and Associated General Contractors.

Paul Tente
Paul Tente Associates
Roofing Consultant
Colorado Springs, Colorado

Paul Tente learned the roofing trade from his father who worked in the roofing industry for 52 years. After 9 years of roofing experience, Mr. Tente joined the management of two large commercial roofing contracting firms in Minneapolis, Minnesota, and continued for 17 years in this capacity. In 1972, he moved to Colorado Springs and formed Paul Tente Associates.

He is the author of the manual *Roofing Concepts/Principles* published by Paul Tente Associates. The manual has been used by the General Services Administration, the Corps of Engineers, the National Bureau of Standards, and many other organizations and professionals throughout the United States and Canada. Paul Tente regularly conducts roofing seminars nationwide, including at the University of Wisconsin.

His roofing consulting clients include the General Services Administration, the National Park Service, the National Aeronautics Space Administration, architects, general contractors, and corporate owners of properties.

James Warner
Consulting Engineer
Mariposa, California

James Warner is a consulting engineer specializing in structural restoration and solutions to construction problems in general, and has been responsible for directing the development of several hundred projects. He has served on many technical panels, advisory groups, and review boards and appears regularly as a lecturer throughout the world.

Mr. Warner is a registered professional engineer and fellow of the American Society of Civil Engineers and the American Concrete Institute. He is a member of the American Arbitration Association, the Earthquake Engineering Research Institute, and the Society of American Military Engineers. Internationally, he is a member of the Federation Internationale de la Precontrainte, The Concrete Society (Great Britain), the International Association for Earthquake Engineering, and the International Society of Soil Mechanics and Foundation Engineering.

He served as a member of the first U.S. Concrete Leaders Goodwill Delegation to Eastern Europe and the Soviet Union in 1971 and to countries in South America in 1973. He is a former member of the National Science Foundation-sponsored U.S./Japan Cooperative Research Committee on repair and retrofit of structures.

Index